ARITHMETIC SKILLS

A Study Guide
for Success

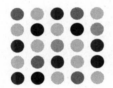

Dr. Daniel M. Seaton

University of Maryland
Eastern Shore

KENDALL/HUNT PUBLISHING COMPANY
4050 Westmark Drive   Dubuque, Iowa 52002

Cover images © 2006 JupiterImages Corporation

Copyright © 2007 by by Daniel M. Seaton

ISBN 13: 978-0-7575-3903-9
ISBN 10: 0-7575-3903-3

Printed in the United States of America
10 9 8 7 6 5 4 3 2 1

Acknowledgments

I would like to thank the staff at Kendall/Hunt Publishing, including Heather Austin and Stefani DeMoss. I would also like to thank Kera Martin and Karen Leung for their editorial assistance with this manuscript.

C O N T E N T S

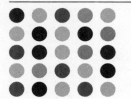

SECTION 1

Numerals: Forms and Place Value

1.1 WHOLE NUMBER PLACE VALUES

In this section you will become familiar with expressing numbers in different forms, understanding place value, and rounding numbers. When given a number, it is important to be able to express it in different ways. For instance, when writing a check, we are required to write the amount out in two ways, standard form and word form. Numbers in standard form are called numerals, Such as 2,539. Written form is actually writing the number out in words: two thousand, five hundred thirty-nine. Another way of expressing a number is in expanded form, Such as $2 \times 1,000 + 5 \times 100 + 3 \times 10 + 9 \times 1$.

To understand how to write a number in different forms, it is necessary to review what you know about place values.

Place value is determined by the placement of each digit or number. Beginning at the right, the ones place has a value of one. Each subsequent place to the left has a value that increases by a factor of ten.

Millions	Hundred thousand	Ten thousands	Thousands	Hundreds	Ten	Ones
$100,000 \times 10$	$10,000 \times 10$	$1,000 \times 10$	100×10	10×10	10×1	1

For our numeral: 2,539

Millions	Hundred thousands	Ten thousands	Thousands	Hundreds	Tens	Ones
0	0	0	2	5	3	9

Expanded form:

$2 \times 1000 = 2000$ $5 \times 100 = 500$ $3 \times 10 = 30$ $9 \times 1 = 9$

$$2,000 + 500 + 30 + 9 = 2,539$$

In words: Two thousand, five hundred thirty-nine

Let's look at a larger example: 2,875,051

Millions	Hundred thousands	Ten thousands	Thousands	Hundreds	Tens	Ones
2	8	7	5	0	5	1

Expanded form:

$2 \times 1,000,000 = 2,000,000$ $8 \times 100,000 = 800,000$ $7 \times 10,000 = 70,000$ $5 \times 1,000 = 5,000$

$0 \times 100 = 0$ $5 \times 10 = 50$ $1 \times 1 = 1$

$$2,000,000 + 800,000 + 70,000 + 5,000 + 0 + 50 + 1 = 2,857,051$$

Notice that in this example, there is a zero in the hundreds place. We use a zero to hold places that are empty.

In words: Two million, eight hundred seventy-five thousand, fifty-one.

EXAMPLE 1:   Write in numerals: Five million five

What do we know about this number? We know that there are 5 millions, but what about the five; what is that? Because there is nothing after the 5, we know that it is five ones. Because there are six digits before the millions place, we will need zeros to fill any empty spots.

Millions	Hundred thousands	Ten thousands	Thousands	Hundreds	Tens	Ones
5	0	0	0	0	0	5

Answer: 5,000,005

EXAMPLE 2:   Write in numerals: Seven hundred fifty-four thousand, ninety-five

Millions	Hundred thousands	Ten thousands	Thousands	Hundreds	Tens	Ones
0	7	5	4	0	9	5

Answer: 754,095

PRACTICE:

Fill in the missing information

1. Write in numerals: Ninety-five thousand, six hundred fifty-two

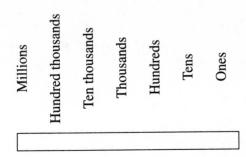

Ninety-five thousand, six hundred fifty-two = _____

2. Write in numerals: Three hundred five thousand, fifty. Do this one without a chart.
   Three hundred five thousand, fifty = _____

1.2 DECIMAL PLACE VALUES

Place values for decimals are very similar to that of whole numbers. Decimals are parts of whole numbers. They also are another way of writing fractions.

You will learn about converting decimals to fractions and vice-versa later in the book.

The objective of this section is to help you understand the place values of decimals. When you see a whole number, there is an implied decimal after it: 5 = 5.00. We see decimals used every day. The best example is using decimals to express a monetary value, such as $2.85 = two dollars, eighty-five cents. The .85 represents a part of the next whole number. What value does each of the numbers following the decimal have?

Place values for decimals:

Ones	.	Tenths	Hundredths	Thousandths	Ten-thousandths	Hundred-thousandths
1	.	$1 \div 10$	$1 \div 100$	$1 \div 1,000$	$1 \div 10,000$	$1 \div 100,000$

EXAMPLE 1:   2.85

Ones	.	Tenths	Hundredths	Thousandths
2	.	8	5	0

EXAMPLE 2:   **0.90437**

Ones	.	Tenths	Hundredths	Thousandths	Ten-thousandths	Hundred-thousandths
0	.	9	0	4	3	7

There are a few important things to remember about decimal place values. First, there is no ones place to the right of the decimal. Second, every place name after the decimal ends with "ths." Third, the place value gets smaller the farther to the right you go from the decimal.

For our example 0.90437:

$0 \times$ *Ones* $= 0$

$9 \times$ *Tenths* $= .9$

$0 \times$ *Hundredths* $= 0$

$4 \times$ *Thousandths* $= .004$

$3 \times$ *Ten-thousandths* $= .0003$

$7 \times$ *Hundred-thousandths* $= .00007$

EXAMPLE 3:   **0.02106, which number is in the thousandths place?**

Ones	.	Tenths	Hundredths	Thousandths	Ten-thousandths	Hundred-thousandths
0	.	0	2	1	0	6

Answer: 1 is in the thousandths place.

EXAMPLE 4:   **5.14975, which place is the 7 in?**

Ones	.	Tenths	Hundredths	Thousandths	Ten-thousandths	Hundred-thousandths
5	.	1	4	9	7	5

Answer: 7 is in the ten-thousandths place.

PRACTICE

1. 256.02356: Which number is in the ten-thousandths place? Fill in the numerals.

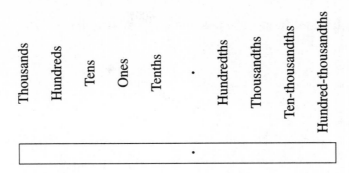

2. What place is each of the underlined digits in?

A. 3.9<u>7</u>5   B. 0.765<u>2</u>

3. Compare. Write >, <, = for each.

A. 0.056 __ 0.560   B. 12.010 __ 12.0010

1.3 WRITING DECIMALS IN WORDS AND STANDARD FORM

Like whole numbers, decimals can also be written in words. For example, two hundredths is 0.02:

Ones	.	Tenths	Hundredths
0	.	0	2

What we say depends on at which place the numeral ends. For 0.5432 we say five thousand four hundred thirty-two ten-thousandths:

Ones	.	Tenths	Hundredths	Thousandths	Ten-thousandths
0	.	5	4	3	2

When there are numerals to the left of the decimal, we say "and" to separate the whole numbers from the decimals. For 12.85 we say twelve *and* eighty-five hundredths. For 56.846 we say fifty-six *and* eight hundred forty-six thousandths.

EXAMPLE 1:   **Write in words, 0.2305**

Answer: two thousand three hundred five ten-thousandths

EXAMPLE 2:   **Write in standard form, sixty-five and eight thousandths**

Answer: 65.008

PRACTICE:

Write each decimal in words.

    1. 0.56   2. 8.025

Write each number in standard form.

    3. Ten and thirty-two thousandths

    4. Seventy-eight ten-thousandths

    5. Which underlined digit has the value five thousandths?

    A. <u>5</u>325.42   B. 7.<u>5</u>227
    C. 0.250<u>5</u>   D. 2.00<u>5</u>

1.4 ROUNDING WHOLE NUMBERS AND DECIMALS

Now that you know place values for whole numbers and decimals, we can begin rounding.
Rounding is used to make it easier to think about very large or very small numbers. You also round to get an approximate value. Rounding almost never results in an accurate figure—only an approximation.

BASIC RULES OF ROUNDING:

Round the tens places to the nearest ten; round up for digits 5 through 9 and round down for digits 0 through 4.

    Example: Round 54 to the nearest ten.
    Answer: 5<u>4</u> rounds down to 50

    Example: Round 77 to the nearest ten.
    Answer: 7<u>7</u> rounds up to 80

Round the hundreds place to the nearest hundred; round up if the number ends in 50 or more and round down for 0 through 49.

    Example: Round 568 to the nearest hundred
    Answer: 5<u>68</u> rounds up to 600

    Example: Round 805 to the nearest hundred
    Answer: 8<u>05</u> rounds down to 800

Likewise, if asked to round to the nearest thousand, round up for numbers 500 or more and round down for numbers 499 and under.

Most of the time, you will be asked to round to a certain place name or place number. For example, *round off* 1.90437 to the hundredths place. To *round off* means to stop at the specified point, in this case the hundredths place. To do this you start by first finding the hundredths place in this decimal.

EXAMPLE 1:   1.90|437

The line separates the hundredths and thousandths place. To round to the hundredths place, look at the number in the thousandths place. If this number is 0–4, just drop it off. If the number is 5 or above, increase the number in the hundredths place by one. 1.90|437 rounded off is 1.90.

EXAMPLE 2:   Round off 0.8594 to the nearest tenths place

Answer: 0.9

EXAMPLE 3:   Round off 0.3321 to the nearest thousandths place

Answer: 0.332

Caution: When faced with a numeral that has whole numbers and decimals, be careful to round to the correct place. For example: Round off 654.019 to the hundreds place. (Not *hundredths* place!) Compare right and wrong answers.

Right: 700   **Wrong:** 654.02

There's a big difference between the two!

Rounding to a place number is similar. The difference is that you have to count to find the correct place.

EXAMPLE 4:   Round off 0.8572165 to four digits

Start by counting off four digits: 0.8572|165. Look at the fifth digit and round up or down from there.

Answer: 0.8572165 rounded to four digits is 0.8572.

PRACTICE:

1.   Round 12.875 to the tenths place. Underline the digit(s) used to round.

2.   Round 765.225 to the hundredths place. Underline the digit(s) used to round.

3.   Round 549.006 to the hundreds place. Underline the digit(s) used to round.

EXERCISES

Section 1.1 Whole number place values

For exercises 1-10, write each in numerals.

1. Seventeen thousand, four hundred thirty

2. Six hundred twenty-one thousand, five

3. Four hundred million forty-one thousand

4. Seven hundred ninety-two thousand, twenty

5. Six million fifty-two

6. Five hundred fifty thousand, two

7. Eighty-two thousand seventy-four

8. Three million four hundred fifty thousand

9. Four hundred thousand seventy-eight

10. Sixteen million twenty-four

Section 1.2 Decimal place values

For exercises 11-15, what place is each of the underlined digits in?

11. 34.1<u>2</u>

12. 0.02<u>3</u>

13. 0.800<u>34</u>

14. 4<u>3</u>5.0038

15. 0.<u>0</u>063

Compare. Write >, <, or = for each.

16. 8.001___ 8.01

17. 3.0682___ 3.0628

Section 1.3 Decimal forms

For exercises 18-22, write each decimal in words

18. 0.52

19. 0.038

20. 0.408

21. 2.014

22. 0.5900

For exercises 23-27, write each decimal in standard form.

23. Two hundredths

24. Seventy-five thousandths

25. Four hundred sixteen thousandths

26. Seventeen and eighty thousandths

27. Ninety ten-thousandths

Section 1.4 Rounding

For exercises 28-30, round to the place of the underlined digit.

28. <u>3</u>.099

29. 0.2<u>6</u>89

30. 9.<u>9</u>745

For exercises 31-35, round off.

31. 1.62713 to the tenths place

32. 1.90437 to the hundredths place

33. 1.60087 to the thousandths place

34. 1.3986 to the ones place

35. 1.35868 to the tenths place

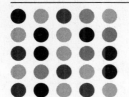

S E C T I O N 2

Adding and Subtracting Whole Numbers and Decimals

2.1 ADDING WHOLE NUMBERS

The purpose of this section is to teach you how to add whole numbers. This section will prepare you for question 1 on the Arithmetic Skills Test. When given numbers to add together it is important to first know the **place value** of each digit within each number. We covered place values in Section 1.

 If you don't remember, then right now you should go back to Section 1 and review place values of whole numbers.

Now that you've reviewed place values, let's get started! Besides the Arithmetic Skills Test, you will also have to add numbers of different sizes on a daily basis, so it is a very important concept to understand. The first step in adding whole numbers is to write the numbers without the comma vertically putting one number above the other. When lining the numbers up you should make sure that each place value in each number is in line with one another.

For the numbers 64,839 and 1,398, let's add them together.

STEP 1:

STEP 1: Identify the place value of each number and align the numbers vertically.
***Many students prefer to put the larger number on top.**

larger number   6 4 8 3 9
smaller number   + 1 3 9 8   =

6 **ten thousands** + 4 **thousands** + 8 **hundreds** + 3 **tens** + 9 **ones**
+ 1 **thousands** + 3 **hundreds** + 9 **tens** + 8 **ones**

STEP 2:   6 4 8 3 9
       + 1 3 9 8

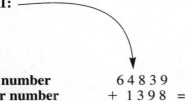

This is the ones column.

STEP 2: Add each of the place values in the top number with same place value in the bottom number. Start with the ones column first.

That means you add: 9 + 8 = 17   ←——   you add the ones column first
3 + 9 = 12   ←——   you add the tens column second
8 + 3 = 11   ←——   you add the hundreds column third
4 + 1 = 5   ←——   you add the thousands column fourth
6 + 0 = 6   ←——   you add the ten thousands column fifth

You will continue to add numbers together until there are NO place values left.

Important: Whenever the sum of a column is a two-digit number, for example $\underline{10}$, then the left number must be carried and added to the NEXT column.

In our example 9 + 8 = 17, and 17 is a two-digit number, so we will take the 1 in the number 17 and carry it to the next column.

$$\begin{array}{r} 1 \\ 6\,4\,8\,3\,9 \\ +\ \ 1\,3\,9\,8 \\ \hline 7 \end{array}$$

Now we will add up each number in this column. (1 + 3 + 9 = 13)

Since 1 + 3 + 9 = 13, we will take the 1 in the number 13 and carry it to the next column.

$$\begin{array}{r} 1\,1 \\ 6\,4\,8\,3\,9 \\ +\ \ 1\,3\,9\,8 \\ \hline 3\,7 \end{array}$$

Now we will add the hundreds column together: 1 + 8 + 3 = 12. Since 12 is a two-digit number, we will take the 1 in the number 12 and carry it to the next column.

$$\begin{array}{r} 1\,1\,1 \\ 6\,4\,8\,3\,9 \\ +\ \ 1\,3\,9\,8 \\ \hline 2\,3\,7 \end{array}$$

Next add the thousands column together: 1 + 4 + 1 = 6   ←—— This isn't a two-digit number so we don't have to carry anything to the next column.

$$\begin{array}{r} 1\,1\,1 \\ 6\,4\,8\,3\,9 \\ +\ \ 1\,3\,9\,8 \\ \hline 6\,2\,3\,7 \end{array}$$

Now our last step is to add the ten-thousands column together: 6 + 0 = 6

$$\begin{array}{r} 1\,1\,1 \\ 6\,4\,8\,3\,9 \\ +\ \ 1\,3\,9\,8 \\ \hline 6\,6\,2\,3\,7 \end{array}$$

Therefore, 6 4 8 3 9 + 1 3 9 8 = 6 6, 2 3 7.

EXAMPLE 1:   Add 595 + 2,042 + 147

Fill in the blanks.

STEP 1:

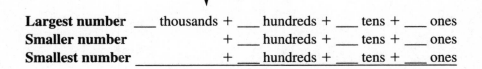

Largest number	___ thousands +	___ hundreds +	___ tens +	___ ones	
Smaller number	+	___ hundreds +	___ tens +	___ ones	
Smallest number	+	___ hundreds +	___ tens +	___ ones	

$$\text{ANSWER} \rightarrow \quad \begin{array}{r} 2042 \\ 595 \\ +\ \underline{14} \end{array}$$

STEP 2:

$$\begin{array}{r} 1 \\ 2042 \\ 595 \\ +\ \underline{147} \\ 4 \end{array}$$

STEP 2: Starting with the ones column add all the numbers that appear in that specific column.

STEP 3:

$$\begin{array}{r} 11 \\ 2042 \\ 595 \\ +\ \underline{147} \\ 84 \end{array}$$

STEP 3: Now add all the numbers that appear in the tens column.

STEP 4:

$$\begin{array}{r} 11 \\ 2042 \\ 595 \\ +\ \underline{147} \\ 784 \end{array}$$

STEP 4: Now add all the numbers that appear in the hundreds column.

STEP 5:

$$\begin{array}{r} 11 \\ 2042 \\ 595 \\ +\ \underline{147} \\ 2784 \end{array}$$

STEP 5: Now add all the numbers that appear in the thousands column.

The answer to 2,042 + 595 + 147 is <u>2,784</u>.

EXAMPLE 2:   Add 6,378 + 890 + 37

STEP 1: Identify the place values of each number. Then align the numbers vertically, putting the largest number first, the smaller number second, and the smallest number last.

STEP 2:

$$\begin{array}{r} 1 \\ 6378 \\ 890 \\ +\ \underline{37} \\ 5 \end{array}$$

STEP 2: Starting with the ones column add all the numbers that appear in that specific column.

STEP 3:

```
      21
    6378
     890
  +   37
  ------
      05
```

STEP 3: Now add all the numbers that appear in the tens column.

STEP 4:

```
      21
    6378
     890
  +   37
  ------
     305
```

STEP 4: Now add all the numbers that appear in the hundreds column.

STEP 5:

```
     121
    6378
     890
  +   37
  ------
    7305
```

STEP 5: Now add all the numbers that appear in the thousands column.

The answer to 6,378 + 890 + 37 is <u>7,305.</u>

PRACTICE:

1. Add the numbers 8,907 + 3,837 + 45 together.

    Larger number
    Smaller number
    Smallest number   + _____

2. Add the numbers 389,337 + 3,738 together.

    Larger number
    Smaller number   + _____

3. Add the numbers 37 + 849 + 23 + 38,930 together.

    Larger number
    Second largest
    Smaller number
    Smallest number   + _____

4. Add the numbers 6 + 390 + 937 together.

    Larger number
    Smaller number
    Smallest number   + _____

2.2   SUBTRACTING WHOLE NUMBERS

The purpose of this section is to teach you how to subtract whole numbers. This section will prepare you for question 2 on the Arithmetic Skills Test. When given numbers to subtract from one another it is important to first know the **place value** of each digit within each number. We covered place values in Section 1.

 **If you don't remember, then right now you should go back to Section 1 and review place values of whole numbers.**

Now that you've reviewed place values, let's get started! Besides the Arithmetic Skills Test, you will also have to subtract numbers of different sizes on a daily basis, so it is very important know how to compute and understand these numbers. The first step in subtracting numbers is to write the numbers without the comma vertically putting one number above the other. When lining the numbers up you should make sure that each place value in each number is in the same column as the place value of the other number.

For the numbers 64,839 and 1,398, let's subtract 1,398 from 64,839.

This word means *take something away*.
So take 1,398 **away from** 64,839 **by using subtraction.**

STEP 1:

STEP 1: Identify the place value of each number and align the numbers vertically. **Make sure to put the larger number on top.*

larger number   6 4 8 3 9   6 **ten thousands** − 4 **thousands** − 8 **hundreds** − 3 **tens** − 9 **ones**
smaller number   − 1 3 9 8 =   − 1 **thousands** − 3 **hundreds** − 9 **tens** − 8 **ones**

STEP 2:   6 4 8 3 9
          + 1 3 9 8

This is the ones column.

STEP 2: Subtract each of the place values (digits) in the bottom number from the same place value (digit) in the top number. Start with the ones column first.

That means we subtract: $9 - 8 = 1$ ⟵   You subtract the ones column first.

$3 - 9 = 4$ ⟵   You subtract the tens column second. **YOU WILL HAVE TO BORROW FROM THE PREVIOUS COLUMN SO THAT YOU DON'T GET A NEGATIVE.**

$8 - 3 = 5$ ⟵   You subtract the hundreds column third.
$4 - 1 = 3$ ⟵   You subtract the thousands column fourth.
$6 - 0 = 6$ ⟵   You subtract the ten-thousands column fifth.

You will continue to subtract numbers until there are NO place values (digits) left.

Important: Whenever the top number in a column is SMALLER than the bottom number in the column you must use a technique called **BORROWING.**

In our example $3 - 9 =$ _____, 3 is SMALLER than 9 so we must borrow 1 unit from the top number in the column to the left of our original column and add 10 units to the top number in our original column.

$$
\begin{array}{r}
6\,4\,8\,3\,9 \\
-\ \ 1\,3\,9\,8 \\
\hline
\uparrow\ 1
\end{array}
$$

Now we will subtract the numbers in this column.

Since 3 is smaller than 9 we will borrow 1 unit from the 8 in the column to the left, making it now a 7. We then add 10 units to the 3. After performing the borrowing method we can now subtract 9 from 13.

$$
\begin{array}{r}
7 \\
\ \ {}^{1} \\
6\,4\,\not{8}\,3\,9 \\
-\ \ 1\,3\,9\,8 \\
\hline
4\,1
\end{array}
$$

Now we will subtract the numbers in the hundreds column: $7 - 3 = 4$.

$$
\begin{array}{r}
7\ {}_{1} \\
6\,4\,\not{8}\,3\,9 \\
-\ \ 1\,3\,9\,8 \\
\hline
4\,4\,1
\end{array}
$$

Next subtract the numbers in the thousands column: $4 - 1 = 3$

$$
\begin{array}{r}
7_{1} \\
6\,4\,\not{8}\,3\,9 \\
-\ \ 1\,3\,9\,8 \\
\hline
3\,4\,4\,1
\end{array}
$$

Now our last step is to subtract the numbers in the ten-thousands column: $6 - 0 = 6$

$$
\begin{array}{r}
7_{1} \\
6\,4\,\not{8}\,3\,9 \\
-\ \ 1\,3\,9\,8 \\
\hline
6\,3\,4\,4\,1
\end{array}
$$

Therefore, $6\,4\,8\,3\,9 - 1\,3\,9\,8 = 6\,3,4\,4\,1$.

EXAMPLE 1:   Subtract 595 from 2,042

Fill in the blanks.

STEP 1:

STEP 1: Identify the place values of each number. Them align the numbers vertically, putting the number following "from" on top and the number found before "from" on the bottom.

Number that follows *from*   ___ thousands + ___ hundreds + ___ tens + ___ ones

Number found before *from*   − ___ hundreds − ___ tens − ___ ones

ANSWER

$$
\begin{array}{r}
2\,0\,4\,2 \\
-\ \ 5\,9\,5
\end{array}
$$

STEP 2:

$$\overset{3}{\cancel{2}}\overset{1}{0}\cancel{4}2$$
$$-595$$
$$\overline{7}$$

STEP 2: Starting with the ones place value, subtract the numbers that appear in that column. **NOTE: 2 is smaller than 5 so you will have to borrow.**

STEP 3:

$$\overset{1}{\cancel{2}}\overset{3}{0}\overset{1}{\cancel{4}}2$$
$$-595$$
$$\overline{7}$$

STEP 3: Subtract the tens column. Now here the 3 is smaller than the 9 so we will have to borrow again. **BUT WAIT!** We can't borrow from 0. So we go the next column to the left, which is 2. After we borrow 1 unit from 2, then 10 units are added to the 0.

STEP 4:

$$\overset{1}{1}\overset{1}{9}\overset{3}{\cancel{0}}\overset{1}{\cancel{4}}2$$
$$-595$$
$$\overline{47}$$

STEP 4: We still need to make the 3 larger so we will borrow 1 unit from the 10 and add 10 units to the 3. After that step we can now subtract normally.

STEP 5:

$$\overset{1}{1}\overset{1}{9}\overset{3}{\cancel{0}}\overset{1}{\cancel{4}}2$$
$$-595$$
$$\overline{147}$$

STEP 5: Subtract the hundreds column.

STEP 6:

$$\overset{1}{1}\overset{1}{9}\overset{3}{\cancel{0}}\overset{1}{\cancel{4}}2$$
$$-595$$
$$\overline{1147}$$

STEP 6: Subtract the thousands column.

The answer to 2,042 − 595 is <u>1,447</u>.

EXAMPLE 2:   Subtract 504 from 7873

STEP 1:

STEP 1: Identify the place values of each number. Then align the numbers vertically, putting the number following "from" on top and the number found before "from" on the bottom.

Number that follows *from*   ___ thousands + ___ hundreds + ___ tens + ___ ones

Number found before *from*   − ___ hundreds − ___ tens − ___ ones

ANSWER
$$7873$$
$$-504$$

STEP 2:

$$\overset{6}{78}\cancel{7}3$$
$$-504$$
$$\overline{9}$$

STEP 2: Starting with the ones place value, subtract the numbers that appear in that column. **NOTE: 3 is smaller than 4 so you will have to borrow.**

STEP 3:

$$\overset{6}{78}\cancel{7}3$$
$$-504$$
$$\overline{69}$$

STEP 3: Subtract the tens column.

STEP 4:

$$\overset{6}{78}\cancel{7}3$$
$$-504$$
$$\overline{369}$$

STEP 4: Subtract the hundreds column.

STEP 5:

$$
\begin{array}{r}
6\\
7\,8\,\not{7}\,3\\
-\quad\ \ 5\,0\,4\\
\hline
7\,3\,6\,9
\end{array}
$$

STEP 5: Subtract the thousands column.

The answer to 7,873 − 504 is <u>7,369</u>.

PRACTICE:

1. Subtract 645 from 5737.

    Number that follows *from*
    Number found before *from*   − _____

2. Subtract 138 from 3746.

    Number that follows *from*
    Number found before *from*   − _____

3. Subtract 456 from 45578.

    Number that follows *from*
    Number found before *from*   − _____

4. Subtract 48 from 169

    Number that follows *from*
    Number found before *from*   − _____

2.3   ADDING DECIMALS

The purpose of this section is to teach you how to add decimals together. It will prepare you for question 5 on the Arithmetic Skills Test. Adding decimals together is similar to adding different amounts of money together. Throughout your life you will have to deal with money, which is why it is important that you understand addition on decimals.

Now, when adding decimals it is important to first know the **place values** of each number you are trying to add. We covered place values of decimals in Section 1.

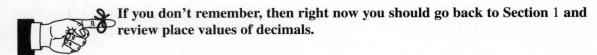 **If you don't remember, then right now you should go back to Section 1 and review place values of decimals.**

Now that you've reviewed place values, let's get started! As noted earlier, besides the Arithmetic Skills Test, you will also have to add decimal numbers of different sizes on a daily basis, so it is very important know how to compute and understand decimals. The first step in adding decimal numbers together is to write the numbers vertically in line with one another (putting one number above the other number). When performing this step make sure to line up the decimal points.

Let's try to add 2.47 and 4.57 together.

STEP 1:

$$
\begin{array}{r}
2.47 \\
+\ 4.57 \\
\hline
. \\
\end{array}
$$

Decimal points
lined up

STEP 1: Write the numbers vertically and be sure to line up the decimal points. Write the decimal point where it will appear in the answer before adding the numbers.
*Lining up the decimal point has the effect of aligning similar place values.

Had to carry
the 1

STEP 2:

$$
\begin{array}{r}
1 \\
2.47 \\
+\ 4.57 \\
\hline
.\ 4 \\
\end{array}
$$

STEP 2: Always start with the number farthest to the right. Therefore, in this example we will first add the hundredths places together. 7 + 7 = 14, and from Section 2.1 we know that when you have a two-digit number you must carry the 1 and add it to the next column.

STEP 3:

$$
\begin{array}{r}
1\ \ 1 \\
2.47 \\
+\ 4.57 \\
\hline
.04 \\
\end{array}
$$

STEP 3: Add the tenths column, being sure to include the 1 that was carried over. Now, 4 + 5 + 1 = 10, which is a two-digit number, so we must carry the 1 and add it to the next column.

STEP 4:

$$
\begin{array}{r}
1\ \ 1 \\
2.47 \\
+\ 4.57 \\
\hline
7.04 \\
\end{array}
$$

STEP 4: Add the ones column, being sure to include the 1 that was carried over.

The answer to 2.47 + 4.57 is 7.04.

EXAMPLE 1:   Add 42.6 + 55. 7 + 62.8

Fill in the blanks.

STEP 1:

$$
\begin{array}{r}
. \\
. \\
+\ . \\
\hline
\end{array}
$$

Write the numbers
correctly in the
space provided

STEP 1: Write the numbers vertically and be sure to line up the decimal points. Write the decimal point where it will appear in the answer before adding the numbers.

STEP 2:

$$
\begin{array}{r}
2 \\
62.8 \\
55.7 \\
+\ 42.6 \\
\hline
1 \\
\end{array}
$$

STEP 2: Always start with the number farthest to the right. Therefore, in this example we will first add the numbers in the tenths column together. 8 + 7 + 6 = 21, and from Section 2.1 we know that when you have a two-digit number you must carry the 2 and add it to the next column.

STEP 3:

$$
\begin{array}{r}
1\ 2 \\
62.8 \\
55.7 \\
+\ 42.6 \\
\hline
1.1 \\
\end{array}
$$

STEP 3: Add the ones column, being sure to include the 2 that was carried over. Now, 2 + 2 + 5 + 2 = 11; this is a two-digit number so we must carry the 1 and add it to the next column.

STEP 4:

```
        1 2
      6 2 . 8
      5 5 . 7
    +  4 2 . 6
      1 6 1 . 1
```

STEP 4: Add the tens column, being sure to include the 1 that was carried over.

The answer to 42.6 + 55. 7 + 62.8 is <u>161.1</u>.

EXAMPLE 2:   Add 123.893 + 1245. 33 + 34.847

Fill in the blanks.

STEP 1:

```
        .
        .
        .
    +   .
        .
```

Write the numbers correctly in the space provided

STEP 1: Write the numbers vertically and be sure to line up the decimal points. Write the decimal point where it will appear in the answer before adding the numbers.

> Adding the zero is optional. It just looks better

STEP 2:

```
              1
      1 2 4 5 . 3 3 0
        1 2 3 . 8 9 3
    +     3 4 . 8 4 7
              .     0
```

STEP 2: Always start with the number farthest to the right. Therefore, in this example we will first add the numbers in thousandths column together. 0 + 3 + 7 = 10, and from Section 2.1 we know that when you have a two-digit number you must carry the 1 and add it to the next column.

STEP 3:

```
            1 1
      1 2 4 5 . 3 3 0
        1 2 3 . 8 9 3
    +     3 4 . 8 4 7
              .   7 0
```

STEP 3: Add the hundreths column, being sure to include the 1 that was carried over. Now, 1 + 3 + 9 + 4 = 17; this is a two-digit number so we must carry the 1 and add it to the next column.

STEP 4:

```
          2 1 1
      1 2 4 5 . 3 3 0
        1 2 3 . 8 9 3
    +     3 4 . 8 4 7
              . 0 7 0
```

STEP 4: Add the tenths column, being sure to include the 1 that was carried over. Now, 1 + 3 + 8 + 8 = 20; this is a two-digit number so we must carry the 2 and add it to the next column.

STEP 5:

```
        1 2 1 1
      1 2 4 5 . 3 3 0
        1 2 3 . 8 9 3
    +     3 4 . 8 4 7
            4 . 0 7 0
```

STEP 5: Add the ones column, being sure to include the 2 that was carried over. Now, 2 + 5 + 3 + 4 = 14, this is a two-digit number so we must carry the 1 and add it to the next column.

STEP 6:

```
    1 1 2  1 1
    1 2 4 5 . 3 3 0
     1 2 3 . 8 9 3
+       3 4 . 8 4 7
       0 4 . 0 7 0
```

STEP 6: Add the tens column, being sure to include the 1 that was carried over. Now, $1 + 4 + 2 + 3 = 10$; this is a two-digit number so we must carry the 1 and add it to the next column.

STEP 7:

```
    1 1 2  1 1
    1 2 4 5 . 3 3 0
     1 2 3 . 8 9 3
+       3 4 . 8 4 7
      4 0 4 . 0 7 0
```

STEP 7: Add the hundreds column, being sure to include the 1 that was carried over.

STEP 8:

```
    1 1 2  1 1
    1 2 4 5 . 3 3 0
     1 2 3 . 8 9 3
+       3 4 . 8 4 7
    1 4 0 4 . 0 7 0
```

STEP 8: Add the thousands column.

The answer to 123.893 + 1245. 33 + 34.847 is <u>1404.070.</u>

PRACTICE:

1. Add: 9.57 + 23.24 + 81.69

2. Add: 5.99 + 0.86

3. Add: 52.83 + 1.95 + .678

4. Add: 87.2 + 1.3 + 15.2

2.4   SUBTRACTING DECIMALS

The purpose of this section is to teach you how to subtract decimal numbers. It will prepare you for question 6 on the Arithmetic Skills Test. When subtracting decimals, one could think about paying money for an item in a store. After you pay for something you will have less money than what you started with. For example, if I have $2.50 and want to buy a popsicle worth $0.25, how much money will I have left? Well, I would have to subtract $0.25 from $2.50, which comes out to be $2.25. Throughout your life you will have to deal with money, which is why it is important that you understand and know how to subtract different amounts of money (decimals).

Now, when subtracting decimals it is important to first know the **place values** of each number you are trying to subtract. We covered place values of decimals in Section 1.

If you don't remember, then right now you should go back to Section 1 **and review place values of decimals.**

If you don't remember how to use the borrowing method, then right now you should go back to Section 2.2 and review subtracting whole numbers.

Now that you've reviewed place values and the borrowing method, let's get started! As noted, besides the Arithmetic Skills Test, you will also have to subtract decimal numbers of different sizes on a daily basis, so it is very important that you are competent in this skill. The first step in subtracting decimal numbers is to write the numbers vertically in line with one another (putting one number above the other number). When performing this step make sure to line up the decimal points.

Let's try to subtract 4.219 from 7.6.

STEP 1:
$$\begin{array}{r} 7.6 \\ -\ 4.219 \\ \hline \end{array}$$

STEP 1: Write the numbers vertically and be sure to line up the decimal points. Write the decimal point where it will appear in the answer before subtracting the numbers.

Decimal points lined up

Don't forget the zeros.

STEP 2:
$$\begin{array}{r} 7.600 \\ -\ 4.219 \\ \hline \end{array}$$

STEP 2: Write the zeros to show hundredths and thousandths place. We want all the numbers to have the same number of places.

STEP 3:
$$\begin{array}{r} 7.6\overset{5\ 9}{\cancel{6}\cancel{0}}0 \\ -\ 4.219 \\ \hline .\quad 1 \end{array}$$

STEP 3: Always start with the number farthest to the right. Therefore, in this example we will first subtract the numbers in the thousandths column. From Section 2.2 we know that when the top number in a column is smaller than the bottom one, we have to use the borrowing method.

STEP 4:
$$\begin{array}{r} 7.6\overset{5\ 9}{\cancel{6}\cancel{0}}0 \\ -\ 4.219 \\ \hline .\ 81 \end{array}$$

STEP 4: Subtract the numbers in the hundredths column.

STEP 5:

$$
\begin{array}{r}
5\,9 \\
7.\,\overset{\cancel{6}\,\overset{1}{}}{}0 \\
-\ 4.2\,1\,9 \\
\hline
.3\,8\,1
\end{array}
$$

STEP 5: Subtract the numbers in the tenths column.

STEP 6:

$$
\begin{array}{r}
5\,9 \\
7.\,\overset{\cancel{6}\,\overset{1}{}}{}0 \\
-\ 4.2\,1\,9 \\
\hline
3.3\,8\,1
\end{array}
$$

STEP 6: Subtract the numbers in the ones column.

The answer to 7.6 − 4.219 is <u>3.381</u>.

EXAMPLE 1: Subtract 52. 40 − 12.75

Fill in the blanks.

STEP 1:

$$
\begin{array}{r}
. \\
-\quad . \\
\hline
.
\end{array}
$$

STEP 1: Write the numbers vertically and be sure to line up the decimal points. Write the decimal point where it will appear in the answer before subtracting the numbers.

STEP 2:

$$
\begin{array}{r}
\overset{3}{}\,\overset{1}{} \\
5\,2.\,\cancel{4}\,0 \\
-\ 1\,2.7\,5 \\
\hline
5
\end{array}
$$

STEP 2: Always start with the number farthest to the right. Therefore, in this example we will first subtract the numbers in the hundredths column. From Section 2.2 we know that when the top number in a column is smaller than the bottom one, we have to use the borrowing method.

STEP 3:

$$
\begin{array}{r}
1\ \overset{1}{3} \\
5\,\cancel{2}.\,\cancel{4}\,0 \\
-\ 1\,2.7\,5 \\
\hline
.6\,5
\end{array}
$$

STEP 3: Subtract the numbers in the tenths column. You will have to use the borrowing method.

STEP 4:

$$
\begin{array}{r}
4\,\overset{1}{1}\ \overset{1}{3} \\
\cancel{5}\,\cancel{2}.\,\cancel{4}\,0 \\
-\ 1\,2.7\,5 \\
\hline
9.6\,5
\end{array}
$$

STEP 4: Subtract the numbers in the ones column. You will have to use the borrowing method.

STEP 5:

$$
\begin{array}{r}
4\,\overset{1}{1}\ \overset{1}{3} \\
\cancel{5}\,\cancel{2}.\,\cancel{4}\,0 \\
-\ 1\,2.7\,5 \\
\hline
3\,9.6\,5
\end{array}
$$

STEP 5: Subtract the numbers in the tens column.

The answer to 52.40 − 12.75 is <u>39.65.</u>

EXAMPLE 2: Subtract 93.71 − 8. 41

Fill in the blanks.

STEP 1:

$$\begin{array}{r} . \\ - \quad . \\ \hline . \end{array}$$

STEP 1: Write the numbers vertically and be sure to line up the decimal points. Write the decimal point where it will appear in the answer before subtracting the numbers.

STEP 2:

$$\begin{array}{r} 9\,3\,.\,7\,1 \\ -\ 5\,1\,.\,4\,1 \\ \hline .\ 0 \end{array}$$

STEP 2: Always start with the number farthest to the right. Therefore, in this example we will first subtract the numbers in the hundredths column.

STEP 3:

$$\begin{array}{r} 9\,3\,.\,7\,1 \\ -\ 5\,1\,.\,4\,1 \\ \hline .\,3\,0 \end{array}$$

STEP 3: Subtract the numbers in the tenths the column.

STEP 4:

$$\begin{array}{r} 9\,3\,.\,7\,1 \\ -\ 5\,1\,.\,4\,1 \\ \hline 2\,.\,3\,0 \end{array}$$

STEP 4: Subtract the numbers in the ones column.

STEP 5:

$$\begin{array}{r} 9\,3\,.\,7\,1 \\ -\ 5\,1\,.\,4\,1 \\ \hline 4\,2\,.\,3\,0 \end{array}$$

STEP 5: Subtract the numbers in the tens column.

The answer to 93.71 − 51.41 is <u>42.30</u>.

PRACTICE:

1. Subtract: 7.29 − 5.36

2. Subtract: 56.95 − 0.869

3. Subtract: 32.83 − 17.95

4. Subtract: 87.2 − 15.2

EXERCISES

Section 2.1 Adding whole numbers

For exercises 1–12, add the whole numbers. Show all work.

1. 135 + 7863 + 127 =

2. 247 + 7682 + 146 =

3. 962 + 2606 + 131 =

4. 691 + 3499 + 13 =

5. 882 + 2115 + 195 =

6. 898 + 1733 + 102 =

7. 106 + 633 + 172 =

8. 4 + 6072 + 90 =

9. 421 + 6089 + 87 =

10. 686 + 9265 + 5 =

11. 333 + 6798 + 11 =

12. 337 + 8360 + 5 =

Section 2.2 Subtracting whole numbers

For exercises 1–12, subtract the whole numbers.

Show all work.

1. 45 from 901 =

2. 296 from 476 =

3. 460 from 895 =

4. 108 from 979 =

5. 627 from 916 =

6. 392 from 1242 =

7. 223 from 980 =

8. 208 from 634 =

9. 135 from 1120 =

10. 589 from 1294 =

11. 188 from 1060 =

12. 171 from 683 =

Section 2.3 Adding decimal numbers

For exercises 1–12, add the decimal numbers.

Show all work.

1. $48.2 + 74 + 36.3 =$

2. $.0223 + .691 + 4.13 =$

3. $.29 + .0952 + .045 =$

4. $.107 + 7.92 + 5.52 =$

5. $56.2 + 5.63 + 6.17 =$

6. $.511 + .0953 + 3.41 =$

7. $98.9 + .0784 + 2.84 =$

8. $.0514 + 1.022 + 35.3 =$

9. $.0617 + 8.82 + .212 =$

10. $.036 + .0313 + .0792 =$

11. $1.26 + .0627 + .0627 =$

12. $6.26 + .0991 + 94 =$

Section 2.4 Subtracting decimal numbers

For exercises 1–12, subtract the decimal numbers.

Show all work.

1. $19.84 - .14 =$

2. $.2915 - .0975 =$

3. $36.08 - 2.78 =$

4. $1.0685 - .0315 =$

5. $12.64 - 10.61 =$

6. $.4788 - .0968 =$

7. $38.742 - 38.5 =$

8. $72.496 - .596 =$

9. $1.3096 - 1.27 =$

10. $77.6703 - .0703 =$

11. $53.2545 - .0545 =$

12. $.9099 - .873 =$

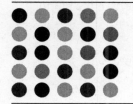

S E C T I O N 3

Multiplication and Division of Whole Numbers and Decimals

3.1 MULTIPLYING WHOLE NUMBERS

In this section, you will review and refine your knowledge of multiplying and dividing whole numbers and decimals.

Multiplication is the repeated addition of a number. For example, $7 \times 4 = 7 + 7 + 7 + 7$. Multiplication is also *commutative,* which means that it doesn't matter which number comes first in a problem, you will get the same answer either way. For our earlier example, $7 \times 4 = 7 + 7 + 7 + 7 = 28$ and $4 \times 7 = 4 + 4 + 4 + 4 + 4 + 4 + 4 = 28$.

To multiply two-two-digit numbers, multiply each term on the top (multiplicand) with each term on the bottom (multiplier), then add the products.

EXAMPLE 1:   **61 $\times$ 70**

STEP 1:
$$\begin{array}{r} 61 \\ \times\ 70 \\ \hline \end{array}$$

STEP 1: Start by rewriting the problem vertically. The top number is called the *multiplicand.* The Bottom number is called The *multiplier.* The final answer is called the *product.*

STEP 2:
$$\begin{array}{r} 61 \\ \times\ 70 \\ \hline 0 \end{array}$$

STEP 2: Multiply 61 by 0.

STEP 3:
$$\begin{array}{r} 61 \\ \times\ 70 \\ \hline 4270 \end{array}$$

STEP 3: Multiple 61 by 7 (product).

EXAMPLE 2:   **17 $\times$ 43**

STEP 1:
$$\begin{array}{r} 43 \\ \times\ 17 \\ \hline \end{array}$$

STEP 1: Rewrite vertically. Remember that multiplication is commutative, so I put 17 on the bottom because I think it's easier to multiply out.

STEP 2:
$$\begin{array}{r} 2 \\ 43 \\ \times\ 17 \\ \hline 301 \end{array}$$

STEP 2: Multiply 43 by 7, *carry* the 2.

STEP 3:

$$\begin{array}{r} 43 \\ \times\ 17 \\ \hline 301 \\ +\ 430 \\ \hline 731 \end{array}$$

STEP 3: Multiply 43 by 1. Put a zero in the empty onces place to hold the spot.

Add products.

 **You can go back to Section 2 if you need to review how to carry digits.**

EXAMPLE 3:   36 × 72

STEP 1:

$$\begin{array}{r} 1 \\ 36 \\ \times\ 72 \\ \hline 72 \end{array}$$

STEP 1: Multiply 36 by 2, *carry* the 1.

STEP 2:

$$\begin{array}{r} 4 \\ 36 \\ \times\ 72 \end{array}$$

STEP 2: Multiply 36 by 7, *carry* the 4.

STEP 3:

$$\begin{array}{r} 72 \\ +\ 2520 \\ \hline 2592 \end{array}$$

STEP 3: Put a zero in to hold the empty ones place.

Add products.

Caution: When adding products, make sure to put them in the correct place!

For example

$$\begin{array}{r} 35 \\ 15 \\ \hline 175 \\ +\ 350 \\ \hline 525 \end{array}$$

Right

$$\begin{array}{r} 35 \\ \times\ 15 \\ \hline 175 \\ +\ 35 \\ \hline 210 \end{array}$$

Wrong

When multiplying, start the product directly under the number you multiplied by.

PRACTICE:

Show work under problem.

1. 50 × 57 =

2. 17 × 25 =

3. $33 \times 47 =$ 　　　　　　　　4. $11 \times 65 =$

5. $27 \times 23 =$

3.2 MULTIPLYING DECIMALS

The process of multiplying decimals is almost identical to multiplying whole numbers. In this section, you will review what you know about multiplying decimals.

Rules of multiplying decimals:

1. Multiply decimal numbers just as you would whole numbers—multiply each of the top numbers with each of the bottom numbers and add the products.
2. Line up the numbers on the right—do not line up the decimals.
3. Place the decimal point in the answer by starting at the right and moving the point the number of places equal to the sum of the decimal places in both numbers being multiplied.

EXAMPLE 1:   **6.7 $\times$ 3.22**

```
      6.7     1 decimal place
  × 3.22      2 decimal places
    134
   1340       Zeros to hold empty places
+ 20100
  21.574      3 decimal places from the right
```

Notice that in this example, we multiplied the numbers the same as we would for whole numbers. The decimal in the answer is three places from the right because there is one place in the first number (multiplicand) and two places in the second (multiplier).

As in whole number multiplication, we again put zeros in to hold the empty spaces. In the third product, there is an empty ones and tens place.

EXAMPLE 2:   **.0091 $\times$ .0103**

```
              *Carry the 2
     2
   .0091      4 places
 × .0103      4 places
   0273
  00000
 009100       Zeros to hold empty places
+0000000
.00009373     8 decimal places from the right
```

This is an example of multiplying decimals containing zeros. The easiest way to do one of these problems is to multiply out the zeros. It takes more time, but it will help you get the correct placement of the decimal. Notice that in this example, the product only contained 7 numbers but the decimal is 8 places from the right. An extra zero had to be added to get the correct decimal placement.

PRACTICE:

Show work under problem.

1. $30.05 \times .05$ 2. $.08 \times .42$

3. 7.19×12.808 4. $.0075 \times .0510$

5. 7.82×1.03

3.3 DIVIDING WHOLE NUMBERS

Division is the inverse, or opposite, of multiplication. $56 \div 7 = 8$ is the inverse of $7 \times 8 = 56$. When we have to divide on paper, we use long division. Long division is the procedure of dividing one number into another. When you see a problem like the first example, you read it as 692 *divided by* 4.

Before we go into the process, there are three definitions that you will need: dividend, divisor, and quotient.

EXAMPLE 1:   **692 ÷ 4**

1. The number to be divided (692) is the *dividend*.
2. The number that you divide by (4) is the *divisor*. This number always comes after the division symbol.
3. The answer is the *quotient*.

STEP 1: $4\overline{)692}$

STEP 1: Start by writing the dividend under the bracket and the divisor on the outside.

STEP 2:
$$\begin{array}{r} 1 \\ 4\overline{)692} \\ -4 \\ \hline 29 \end{array}$$
  $(6 \div 4)$
  $(4 \times 1 = 4)$

STEP 2: Divide 4 into the first number of the dividend that will result in a whole number (6). Place the 1 over the bracket, above the 6. Multiply 4 by 1 and place the answer under the 6, then subtract. Bring down the 9 to make 29.

STEP 3:
$$\begin{array}{r} 17 \\ 4\overline{)692} \\ -4 \\ \hline 29 \\ -28 \\ \hline 12 \end{array}$$

(29 ÷ 4)

(4 × 7 = 28)

STEP 3: Divide 4 into 29 (7) and place next to the 1 in the quotient. Multiply 4 by 7 and place the product under the 29: subtract. Bring down the 2 to make 12.

STEP 4:
$$\begin{array}{r} 173 \\ 4\overline{)692} \\ -4 \\ \hline 29 \\ -28 \\ \hline 12 \\ -12 \\ \hline 00 \end{array}$$

(12 ÷ 4)

(4 × 3 = 12)

STEP 4: Divide 4 into 12 (3) and place next to the 7 in the quotient. Multiply 4 by 3 and place the product under the 12; subtract.

After going through each of the steps required, you can easily check your work by multiplying the quotient by the divisor.

Check:
$$\begin{array}{r} 21 \\ 173 \\ \times 4 \\ \hline 692 \end{array}$$

Let's look at an example with larger numbers.

EXAMPLE 2:   Divide 1752 by 24

This problem is written in words. In standard form, it is written as 1754 ÷ 24. Start by putting the dividend under the bracket and the divisor to the left of the bracket.

STEP 1: $24\overline{)1754}$

STEP 2:
$$\begin{array}{r} 7 \\ 24\overline{)1754} \\ -168 \\ \hline 74 \end{array}$$

(175 ÷ 24)
(24 × 7 = 168)

STEP 3:
$$\begin{array}{r} 73\,r2 \\ 24\overline{)1754} \\ -168 \\ \hline 74 \\ -72 \\ \hline 2 \end{array}$$

(74 ÷ 24)

(24 × 3 = 72)

As you can see, division doesn't always result in a simple answer. The numbers left over are called remainders, so the answer to this problem is expressed as *73 remainder 2*.

$$
\begin{array}{r}
24 \\
\times \quad 73 \\
\hline
72 \\
\end{array}
$$

Check: $+\ 168$

$$
\begin{array}{r}
1752 \\
+ \quad 2 \qquad \text{add remainder}\\
\hline
1754 \\
\end{array}
$$

EXAMPLE 3:   **4955 ÷ 48**

STEP 1:
$$
\begin{array}{r}
1 \\
48\overline{)4955} \\
-48 \\
\hline
15 \\
\end{array}
$$

STEP 2:
$$
\begin{array}{r}
10 \\
48\overline{)4955} \\
-48 \\
\hline
155 \\
\end{array}
$$

STEP 3:
$$
\begin{array}{r}
103r11 \\
48\overline{)4955} \\
-48 \\
\hline
155 \\
-144 \\
\hline
11 \\
\end{array}
$$

Check:
$$
\begin{array}{r}
103 \\
\times \quad 48 \\
\hline
824 \\
+ \ 4120 \\
\hline
4944 \\
+ \quad 11 \qquad \text{add remainder}\\
\hline
4955 \\
\end{array}
$$

PRACTICE:

Show work and checks.

1. 1452 ÷ 66

2. 55$\overline{)34,007}$

3. Divide 580,058 by 58

4. 67,238 ÷ 32

5. Divide 3,450 by 18

3.4 DIVIDING DECIMALS

Dividing decimal numbers is very similar to dividing whole numbers. As with multiplication, there are a couple of things to remember about the decimal point.

To divide one decimal by another, you first have to change the divisor into a whole number. To do this, you have to multiply both the divisor and the dividend with the same multiple of ten (10, 100, 1000, etc.) An easy way to do this is to move the decimal the same number of spaces in each number.

For example: $.8175 \div .18 \rightarrow 81.75 \div 18$.

After multiplying both sides by 100, or moving each decimal to the right twice, we can now successfully complete this problem.

The other thing to remember about the decimal is its placement in the quotient. The decimal in the quotient is directly above the decimal in the dividend.

$$81.75 \div 18 \rightarrow 18\overline{)81.75}$$

EXAMPLE 1:   **.008008 ÷ .077**

STEP 1: $77\overline{)8.008}$

STEP 1: Move the decimal in both the divisor and dividend 3 places [(× 1000)].

STEP 2:
$$\begin{array}{r} .104 \\ 77\overline{)8.008} \\ -77 \\ \hline 308 \\ -308 \\ \hline 00 \end{array}$$

STEP 2: Divide as usual—place the decimal in the quotient in the exact same place as it is in the dividend.

EXAMPLE 2:   **360 ÷ .45**

STEP 1: $45\overline{)36000}$

STEP 1: Move the decimal 2 places (× 100). Notice that we added two zeros onto the dividend.

$$\begin{array}{r} 800 \\ 45{\overline{\smash{)}36000}} \\ -360 \\ \hline 00 \end{array}$$

STEP 2: Divide as usual, and in this case we have a whole number as a quotient.

EXAMPLE 3:   **.6759 ÷ 27**

STEP 1: $27{\overline{\smash{)}.6759}}$

STEP 1: The divisor is already a whole number so the decimal in the dividend stays the same.

STEP 2:

$$\begin{array}{r} .0250 \\ 27{\overline{\smash{)}.6759}} \\ -54 \\ \hline 135 \\ -135 \\ \hline 09 \end{array}$$

STEP 2: This problem is different from the other two examples because the divisor is larger than the dividend and there is a remainder.

When you see a remainder on the end of the quotient in decimal division, it usually means the answer isn't complete. If needed, you can add a zero to the end of the dividend to extend the quotient. This is what Example 3 would look like if we had done that:

$$\begin{array}{r} 0.025033\ldots \\ 27{\overline{\smash{)}.675900}} \\ -54 \\ \hline 135 \\ -135 \\ \hline 090 \\ -81 \\ \hline 90 \\ -81 \\ \hline 9 \end{array}$$

As you can see, this is a repeating decimal. We can keep going, but it doesn't make sense to do so.

PRACTICE:

Show all work.

1. 12.67 ÷ 4.05

2. Divide .0649532 by 7.6

3. 1.75476 ÷ 12.6

4. $15{\overline{\smash{)}9.7051}}$

5. 54.03889 ÷ 6.30

EXERCISES

Section 3.1 Whole number multiplication

Show all work.

1. 29×58

2. 79×18

3. 22×31

4. 97×49

5. 86×74

6. 59×40

7. 37×58

8. 862×7

9. 67×65

10. $8,720 \times 9$

11. 70×23

13. 61×70

12. 972×44

Section 3.2 decimal multiplication

Show all work.

14. 0.0086×3.8

17. 0.109×0.104

15. 0.0091×0.103

18. 0.80×12.01

16. 2.003×0.56

19. 7.4×0.0042

20. 2.014 × 2.110

24. 0.075 × 21.3

21. 0.029 × 0.074

25. 0.005 × 0.99

22. 0.214 × 12.11

26. 2.503 × 0.003

23. 0.0095 × 0.053

27. 6.7 × 0.01

Section 3.3 whole number division

Show all work.

28. 641 ÷ 7

29. 580,058 ÷ 58

30. 5,990 ÷ 23

31. 1452 ÷ 66

32. 1,986 ÷ 64

33. 6,240,208 ÷ 104

34. 45,008 ÷ 90

35. 2,376 ÷ 72

36. 25,755 ÷ 51

37. 2,840,639 ÷ 71

38. 4,908 ÷ 7

39. 11,973 ÷ 39

40. 6,890 ÷ 32

41. 1,752 ÷ 24

Section 3.4 decimal division

Show all work.

42. 0.008008 ÷ 0.077

47. 0.00304 ÷ 0.19

43. 6.16 ÷ 0.7

48. 2.776 ÷ 0.89

44. 0.1099 ÷ 0.008

49. 0.533 ÷ 0.82

45. 0.007695 ÷ 0.0095

50. 0.911 ÷ 0.033

46. 0.9554 ÷ 1.95

51. 0.008181 ÷ 1.01

52. 1.0098 ÷ 3.22

54. 0.099 ÷ 0.03

53. 0.04094 ÷ 8.9

55. 0.0004214 ÷ 0.086

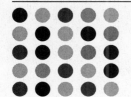

Adding and Subtracting Fractions and Mixed Numbers

4.1 ADDING FRACTIONS

The purpose of this section is to review what you know about adding fractions. This section will prepare you for question 9 on the Arithmetic Skills Test. When given fractions to add together it is important to know the parts of a fraction.

For example, here is a fraction: $\dfrac{1}{4}$

Numerator—**top number**

Denominator—**bottom number**

Another type of fraction is called an **improper fraction**, sometimes referred to as a **top-heavy fraction.** An improper fraction is one in which the numerator is *larger than* the denominator. On the Arithmetic Skills Test you will be asked to change all improper fractions to mixed numbers. **Mixed numbers** are numbers that require a whole number and fraction to write them. For example, $3\dfrac{1}{4}$ is a mixed number. Mixed numbers and improper improper fractions are closely related. Mixed numbers can be changed to improper fractions.

STEP 1: $3\dfrac{1}{4}$

MULTIPLY

ADD THE NUMERATOR

STEP 2: $3\dfrac{1}{4}$

STEP 3: $\dfrac{13}{4}$

STEP 1: Multiply the denominator by the whole number.
$4 \times 3 = 12$.

STEP 2: Add that product to the numerator.
$4 \times 3 = 12 + 1 = 13$.

STEP 3: 13 is now the numerator and the denominator remains the same.

Now, to change from an improper fraction to a mixed number, you do the following:

STEP 1:
$$\begin{array}{r} 3 \\ 4\overline{)13} \\ -12 \\ \hline 1 \end{array}$$

STEP 1: Divide the numerator by the denominator.

STEP 2: $3 + \dfrac{1}{4} = 3\dfrac{1}{4}$

STEP 2: Put the remainder over the divisor. Add this fraction to the quotient. You now have your mixed number.

When it comes to adding fractions, there are some things that you must check for first. For example, are the denominators in each fraction the same or different? If they are the same then we can start adding the fractions. If they are different then we must look for the LCD, **least common denominator**. The LCD is the smallest number that can be the denominator of both fractions.

Let't add $\dfrac{5}{6} + \dfrac{7}{6}$ together.

STEP 1: $\dfrac{5}{6} + \dfrac{7}{6}$ ← Denominators are same number

Only add numerators together

STEP 1: Check to see if the denominators are the same number. In this example they are, so we may add the fractions together without any more steps.

STEP 2: $\dfrac{5 + 7}{6} = \dfrac{13}{6}$

STEP 2: When adding fractions, if the denominators are the same then all we must do is add the numerator and keep the denominator the same.

STEP 3:
$$\begin{array}{r} 2 \\ 6\overline{)13} \\ -12 \\ \hline 1 \end{array} = 2\dfrac{1}{6}$$

STEP 3: Change the improper fraction to a mixed number.

If you are asked to add fractions with different denominators, you must first find the LCD. As stated earlier, the LCD is the smallest number that both the fractions can have. It is the smallest number that can be evenly divided by each denominator.

Add $\dfrac{1}{2} + \dfrac{2}{3}$.

STEP 1: $\dfrac{1}{2} + \dfrac{2}{3}$ ← Denominators are not the same

STEP 1: Check to see if the denominators are the same. In this example they are not, so we must find a least common denominator.

STEP 2: $\overline{6}$ ← LCD

STEP 2: To find the least common denominator you must think about the multiples of each of the denominators. From those multiples find the smallest one that both denominators share. In this example, 6 is our least common denominator.

STEP 3: $\dfrac{1}{2} \otimes \dfrac{3}{3} + \dfrac{2}{3} \otimes \dfrac{2}{2}$

STEP 3: Multiply each fraction's top and bottom by the number that will produce the LCD. In this example $2 \times 3 = 6$, so we multiply the $\dfrac{1}{2}$ by 3 on the top and bottom. The same thing goes for the $\dfrac{2}{3}$, $3 \times 2 = 6$, so we multiply the top and bottom of the fraction by 3.

STEP 4: $\dfrac{3}{6} + \dfrac{4}{6} = \dfrac{7}{6}$

STEP 4: Add the numerators together and keep the denominator the same.

STEP 5: $6\overline{)7} \;\; \begin{array}{r}1\\-6\\\hline 1\end{array} = 1\dfrac{1}{6}$

STEP 5: Change the improper fraction to a mixed number.

EXAMPLE 1: Add: $\dfrac{8}{20} + \dfrac{9}{25}$

STEP 1: $\dfrac{8}{20} + \dfrac{9}{25}$

STEP 1: Check to see if the denominators are the same. In this example they are not, so we must find a least common denominator.

STEP 2: $\overline{100} \leftarrow$ LCD

STEP 2: To find the least common denominator you must think about the multiples of each of the denominators. From those multiples find the smallest one that both denominators share. In this example, 100 is our least common denominator.

STEP 3: $\dfrac{8}{20} \otimes \dfrac{5}{5} + \dfrac{9}{25} \otimes \dfrac{4}{4}$

STEP 3: Multiply each fraction's top and bottom by the number that will produce the LCD. In this example, $20 \times 5 = 100$, so we multiply the $\dfrac{8}{20}$ by 5 on the top and bottom. The same thing goes for the $\dfrac{9}{25}$, $25 \times 4 = 100$, so we multiply the top and bottom of the fraction by 4.

STEP 4: $\dfrac{40}{100} + \dfrac{36}{100} = \dfrac{76}{100}$

STEP 4: Add the numerators together and keep the denominator the same.

STEP 5: $\dfrac{\overset{19}{\cancel{76}}}{\underset{25}{\cancel{100}}} = \dfrac{19}{25}$

STEP 5: Simplify the fraction. To do this, divide the top and bottom by the greatest common multiple.

EXAMPLE 2:   Add $\dfrac{3}{12} + \dfrac{5}{12}$

STEP 1: $\dfrac{3}{12} + \dfrac{5}{12}$

STEP 1: Check to see if the denominators are the same number. In this example they are.

Only add numerators together

STEP 2: $\dfrac{3+5}{12} = \dfrac{8}{12}$

STEP 2: When adding fractions, if the denominators are the same then all we must do is add the numerator and keep the denominator the same.

STEP 3: $\dfrac{\overset{2}{\cancel{8}}}{\underset{3}{\cancel{12}}} = \dfrac{2}{3}$

STEP 3: Simplify the fraction. To do this, divide the top and bottom by the greatest common multiple.

PRACTICE

1. $\dfrac{5}{6} + \dfrac{1}{4} =$

2. $\dfrac{1}{5} + \dfrac{3}{10} =$

3. $\dfrac{31}{50} + \dfrac{9}{50} =$

4. $\dfrac{1}{2} + \dfrac{11}{16} =$

4.2 SUBTRACTING FRACTIONS

The purpose of this section is to teach you how to subtract fractions. This section will prepare you for question 10 on the Arithmetic Skills Test. When given fractions to subtract, it is important to know all the different parts of a fraction.

For example, here is a fraction: $\dfrac{1}{4}$ Numerator—**top number**

Denominator—**bottom number**

As noted, another type of fraction is called an **improper fraction**, sometimes referred to as a **top-heavy fraction.** An improper fraction is one in which the numerator is *larger than* the denominator. On the Arithmetic Skills Test you will be asked to change all improper fractions to mixed numbers. **Mixed numbers** are numbers that require a whole number and fraction to write them. For example, $3\frac{1}{4}$ is a mixed number.

Mixed numbers and improper fractions are closely related. To change a mixed number to an improper fraction you do the following:

STEP 1: $3\frac{1}{4}$ MULTIPLY

STEP 1: Multiply the denominator by the whole number.
$4 \times 3 = 12$.

STEP 2: $3\frac{1}{4}$ ADD THE NUMERATOR

STEP 2: Add that product to the numerator.
$4 \times 3 = 12 + 1 = 13$.

STEP 3: $\frac{13}{4}$

STEP 3: 13 is now the numerator and the denominator remains the same.

Now, to change from an improper fraction to a mixed number, you do the following:

STEP 1:
$$4)\overline{13}$$
$$\frac{-12}{1}$$
with quotient 3

STEP 1: Divide the numerator by the denominator.

STEP 2: $3 + \frac{1}{4} = 3\frac{1}{4}$

STEP 2: Take the remainder and put it over the divisor. Add this fraction to the quotient. You now have your mixed number.

When it comes to subtracting fractions, there are some things that you must check for first. For example, are the denominators in each fraction the same or different? If they are the same then we can start subtracting the numerators. If they are different then we must look for the LCD—**least common denominator.** The LCD is the smallest number that can be the denominator of both fractions.

Let's subtract $\frac{7}{6} - \frac{5}{6}$ together.

STEP 1: $\frac{7}{6} - \frac{5}{6}$ ← Denominators are same number

STEP 1: Check to see if the denominators are the same number. In this example they are the same, so we may subtract the fractions without any more steps.

Only subtract numerators

STEP 2: $\frac{7 - 5}{6} = \frac{2}{6}$

STEP 2: When subtracting fractions, if the denominators are the same then all you have to do is subtract the numbers in the numerator and keep the denominator the same.

STEP 3: $\dfrac{\overset{1}{\cancel{2}}}{\underset{3}{\cancel{6}}} = \dfrac{1}{3}$

STEP 3: Simplify the fraction. To do this, divide the top and bottom by the greatest common multiple.

If you are asked to add fractions with different denominators, you must first find the LCD. As stated earlier, the LCD is the smallest number that both the fractions can have. It is the smallest number that can be evenly divided by each denominator.

Subtract $\dfrac{2}{3} - \dfrac{1}{2}$.

STEP 1: $\dfrac{2}{3} - \dfrac{1}{2}$ Denominators are not the same

STEP 1: Check to see if the denominators are the same. In this example they are not, so we must find a least common denominator.

STEP 2: $\overline{6}$ ← LCD

STEP 2: To find the least common denominator you must think about the multiples of each of the denominators. From those multiples find the smallest one that both denominators share. In this example, 6 is our least common denominator.

STEP 3: $\dfrac{2}{3} \otimes \dfrac{2}{2} - \dfrac{1}{2} \otimes \dfrac{3}{3}$

STEP 3: Multiply each fraction's top and bottom by the number that will produce the LCD. In this example $2 \times 3 = 6$, so we multiply the $\dfrac{1}{2}$ by 3 on the top and bottom. The same thing goes for the $\dfrac{2}{3}$, $3 \times 2 = 6$, so we multiply the top and bottom of the fraction by 3.

STEP 4: $\dfrac{4}{6} - \dfrac{3}{6} = \dfrac{1}{6}$

STEP 4: Subtract the numbers in the numerator and keep the denominator the same.

EXAMPLE 1:   Subtract $\dfrac{8}{20} - \dfrac{9}{25}$

STEP 1: $\dfrac{8}{20} - \dfrac{9}{25}$

STEP 1: Check to see if the denominators are the same. In this example they are not, so we must find a least common denominator.

STEP 2: $\overline{100}$ ↖ LCD

STEP 2: To find the least common denominator you must think about the multiples of each of the denominators. From those multiples find the smallest one that both denominators share. In this example, 100 is our least common denominator.

STEP 3: $\dfrac{8}{20} \otimes \dfrac{5}{5} - \dfrac{9}{25} \otimes \dfrac{4}{4}$

STEP 3: Multiply each fraction's top and bottom by the number that will produce the LCD. In this example, $20 \times 5 = 100$, so we multiply the $\dfrac{8}{20}$ by 5 on the top and bottom. The same thing goes for the $\dfrac{9}{25}$, $25 \times 4 = 100$, so we multiply the top and bottom of the fraction by 4.

STEP 4: $\dfrac{40}{100} - \dfrac{36}{100} = \dfrac{4}{100}$

STEP 4: Subtract the numbers in the numerator and keep the denominator the same.

STEP 5: $\dfrac{\overset{1}{\cancel{4}}}{\underset{25}{\cancel{100}}} = \dfrac{1}{25}$

STEP 5: Simplify the fraction. To do this, divide the top and bottom by the greatest common multiple.

EXAMPLE 2:   Subtract $\dfrac{5}{12} - \dfrac{3}{12}$

STEP 1: $\dfrac{5}{12} - \dfrac{3}{12}$

STEP 1: Check to see if the denominators are the same number. In this example they are, so we may subtract the fractions without any more steps.

Only subtract numerators

STEP 2: $\dfrac{5 - 3}{12} = \dfrac{2}{12}$

STEP 2: When subtracting fractions, if the denominators are the same then all you must do is subtract the numbers in the numerator and keep the denominator the same.

STEP 3: $\dfrac{\overset{1}{\cancel{2}}}{\underset{6}{\cancel{12}}} = \dfrac{1}{6}$

STEP 3: Simplify the fraction. To do this, divide the top and bottom by the greatest common multiple.

PRACTICE:

1. $\dfrac{2}{3} - \dfrac{5}{12} =$

2. $\dfrac{3}{8} - \dfrac{1}{3} =$

3. $\dfrac{4}{5} - \dfrac{2}{3} =$

4. $\dfrac{5}{7} - \dfrac{1}{3} =$

4.3 ADDING MIXED NUMBERS

The purpose of this section is to teach you how to add mixed numbers. This section will prepare you for question 13 on the Arithmetic Skills Test. Adding mixed numbers is a combination of two things you have already learned. You learned how to add whole numbers in Section 2.1, and you learned how to add fractions in Section 4.1. Thus, when you combine both of these processes you have added a mixed number.

Therefore, it is important that you understand and know how to add whole numbers and fractions. We covered whole number addition in Section 2.1 and fraction addition in Section 4.1.

 **If you don't remember, then right now you should go back to Sections 2.1 and 4.1 and review whole number and fraction addition.**

Now that you've reviewed Sections 2.1 and 4.1, let's get started! Let's start with the example $2\dfrac{1}{5} + 3\dfrac{3}{5}$.

Whole number sum

STEP 1: $2 + 3 = 5$

STEP 1: Add the whole numbers together.

STEP 2: $\dfrac{1}{5} + \dfrac{3}{5} = \dfrac{1+3}{5} = \dfrac{4}{5}$

STEP 2: Add the fractions together following the rules and steps you learned in Section 4.1.

Fraction sum

STEP 3: $5 + \dfrac{4}{5} = 5\dfrac{4}{5}$

STEP 3: Add the whole number sum and fraction sum together. This ultimately gives you your mixed number.

The answer to $2\dfrac{1}{5} + 3\dfrac{3}{5}$ is $5\dfrac{4}{5}$

EXAMPLE 1:   Add $6\frac{1}{2} + 2\frac{1}{3}$.

STEP 1: $6 + 2 = 8$

STEP 1: Add the whole numbers together.

STEP 2: $\dfrac{1}{2} + \dfrac{1}{3} = \dfrac{3 + 2}{6} = \dfrac{5}{6}$

STEP 2: Add the fractions together following the rules and steps you learned in Section 4.1. In this example you must find a LCD and then add the numerators after you have changed each fraction.

STEP 3: $8 + \dfrac{5}{6} = 8\dfrac{5}{6}$

STEP 3: Add the whole number sum and fraction sum together. This ultimately gives you your mixed number.

The answer to $6\dfrac{1}{2} + 2\dfrac{1}{3}$ is $8\dfrac{5}{6}$.

EXAMPLE 2:   Add $3\frac{2}{3} + 4\frac{1}{2}$

STEP 1: $3 + 4 = 7$

STEP 1: Add the whole numbers together.

STEP 2: $\dfrac{2}{3} + \dfrac{1}{2} = \dfrac{4 + 3}{6} = \dfrac{7}{6} = 1\dfrac{1}{6}$

STEP 2: Add the fractions together following the rules and steps you learned in Section 4.1. In this example you must find a LCD and then add the numerators after you have changed each fraction. **If you have an improper fraction convert it to a mixed number.**

STEP 3: $7 + 1\dfrac{1}{6} = 8\dfrac{1}{6}$

STEP 3: Add the whole number sum and fraction sum together. This ultimately gives you your mixed number.

The answer to $3\dfrac{2}{3} + 4\dfrac{1}{2}$ is $8\dfrac{1}{6}$

PRACTICE:

1. $4\dfrac{2}{3} + 9\dfrac{5}{6} =$

2. $3\dfrac{1}{2} + 4\dfrac{1}{4} =$

3. $9\dfrac{1}{10} + 3\dfrac{1}{2} =$

4. $2\dfrac{7}{8} + 1\dfrac{1}{2} =$

EXERCISES

Section 4.1 Adding fractions

For exercises 1–12, add the fractions. Show all work.

1. 4/14 + 32/35 =

2. 11/14 + 14/21 =

3. 18/33 + 24/55 =

4. 27/35 + 4/14 =

5. 12/55 + 15/22 =

6. 28/55 + 49/187 =

7. 5/14 + 7/28 =

8. 4/7 + 20/21 =

9. 1/4 + 2/6 =

10. 4/22 + 21/55 =

11. 1/11 + 21/44 =

12. 1/2 + 4/6 =

Section 4.2 Subtracting fractions

For exercises 1–12, subtract the fractions. Show all work.

1. $8/10 - 3/6 =$

2. $30/35 - 17/25 =$

3. $8/15 - 1/9 =$

4. $12/14 - 2/6 =$

5. $3/15 - 3/25 =$

6. $7/21 - 5/35 =$

7. $16/33 - 19/77 =$

8. $5/21 - 2/9 =$

9. $14/15 - 25/35 =$

10. $14/15 - 5/21 =$

11. $7/21 - 6/35$

12. $12/15 - 4/21$

Section 4.3 Adding mixed numbers

For exercises 1–12, add the mixed numbers. Show all work.

1. $6^{39/49} + 8^{2/21} =$

2. $1^{10/21} + 5^{18/49} =$

3. $7^{13/25} + 9^{15/35} =$

4. $6^{8/55} + 8^{21/25} =$

5. $5^{2/4} + 1^{3/4} =$

6. $4^{2/10} + 9^{2/14} =$

7. $8^{2/3} + 8^{1/3} =$

8. $8^{11/22} + 2^{4/22} =$

9. $1^{8/22} + 8^{8/14} =$

10. $3^{1/10} + 7^{9/35} =$

11. $8^{2/33} + 8^{8/9} =$

12. $1^{3/5} + 2^{3/7} =$

Section 4.4 Subtracting mixed numbers

For exercises 1–12, subtract the mixed numbers. Show all work.

1. $6^{1/2} - 2^{1/4} =$

4. $8^{3/8} - 3^{1/64} =$

2. $4^{11/24} - 1^{5/72} =$

5. $12^{2/3} - 5^{1/6}$

3. $2^{6/13} - 1^{12/26} =$

6. $4^{1/4} - 1^{3/16} =$

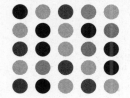

5

Multiplication and Division of Fractions and Mixed Numbers

5.1 MULTIPLYING FRACTIONS

Fractions are another way of writing parts of whole numbers. You will see how to change fractions into decimals and vice-versa in Section 6.

Most people use decimals daily and are comfortable doing so. Fractions, on the other hand, are used less often. We use fractions in many activities, such as cooking and building, so it is important that we know how to add, subtract, multiply, and divide them.

This section will help you recall what you learned in grade school about multiplying and dividing fractions.

Before we begin multiplying and dividing, let's recall some important facts about fractions:

- The top half of a fraction is called the **numerator.**
- The bottom half of a fraction is called the **denominator.**
- To **simplify** means to put the fraction in its lowest terms. To simplify, find the GCF (**greatest common factor**) of the fraction. The GCF of a fraction is the largest number that can divide both the numerator and denominator. For example, $\frac{15}{21}$ has a GCF of 3: $\frac{15}{21} \div \frac{3}{3} = \frac{5}{7}$.

- The **reciprocal** of a fraction is the numerator and denominator flipped around. Example: $\frac{2}{3}$ is the reciprocal of $\frac{3}{2}$.

Multiplying fractions is a relatively simple process. The steps are:

- Simplify the fractions to be multiplied if they're not already in lowest terms.
- Multiply the numerators together.
- Multiply the denominators together.
- Simplify the new fraction if needed.

57

EXAMPLE 1: $\dfrac{6}{9} \times \dfrac{5}{8}$

STEP 1: $\dfrac{6}{9} \times \dfrac{5}{8}$

$\dfrac{6}{9} \div \dfrac{3}{3} = \dfrac{2}{3}$

STEP 1: Simplify $\dfrac{6}{9}$ by dividing both the numerator and denominator by 3 (GCF).

$\dfrac{5}{8}$ is already in lowest terms.

STEP 2: $\dfrac{2}{3} \times \dfrac{5}{8} = \dfrac{2 \times 5}{3 \times 8}$

$\dfrac{2 \times 5}{3 \times 8} = \dfrac{10}{24}$

STEP 2: Multiply the numerators and denominators together.

STEP 3: $\dfrac{10}{24} \div \dfrac{2}{2} = \dfrac{5}{12}$

STEP 3: Simplify the new fraction by dividing both the numerator and denominator by 2 (GCF).

In this example we could have just multiplied the fractions together and then simplified the new fraction, but the simplification would have been more difficult to do. This is what that would look like: $\dfrac{6}{9} \times \dfrac{5}{8} = \dfrac{30}{72}$. To simplify this fraction would take more work, and there would be a greater chance of getting the answer wrong. This is why it is far better to simplify before you multiply.

EXAMPLE 2: $\dfrac{9}{40} \times \dfrac{15}{36}$

STEP 1: $\dfrac{9}{40} \times \dfrac{15}{36} = \dfrac{9 \times 15}{40 \times 36}$

STEP 1: In the original form, neither the numerator nor denominator can be simplified.

STEP 2: $\dfrac{9 \times 15}{40 \times 36} \div \dfrac{9}{9} = \dfrac{1 \times 15}{40 \times 4}$

$\dfrac{1 \times 15}{40 \times 4} \div \dfrac{5}{5} = \dfrac{1 \times 3}{8 \times 4}$

STEP 2: Once the numerators and denominators are together, you can simplify. First we divide the numerator and the denominator of the fraction by 9 to cancel part of the numerator and simplify 36 in the denominator down to 4. Then we divide the numerator and denominator of fraction by 5 to simplify the numerator down to 3 and the denominator down to 8.

STEP 3: $\dfrac{1 \times 3}{8 \times 4} = \dfrac{3}{32}$

STEP 3: Multiply the simplified fractions.

In this example, neither of the original fractions could be simplified on their own so we created the new fraction with both numerators and both denominators. In this new fraction, we could cancel out a 9 and then a 5 from the numerator and denominator.

EXAMPLE 3: $\dfrac{2}{7} \times 3$

STEP 1: $\dfrac{2}{7} \times \dfrac{3}{1}$

STEP 1: When multiplying a fraction by a whole number, place the whole number over a 1. We do this because $3 \div 1 = 3$

STEP 2: $\dfrac{2}{7} \times \dfrac{3}{1} = \dfrac{6}{7}$

STEP 2: Multiply fractions as usual.

PRACTICE:

Multiply and simplify, if possible. Show all work.

1. $\dfrac{4}{7} \times \dfrac{1}{3}$

2. $\dfrac{13}{16} \times \dfrac{8}{52}$

3. $\dfrac{7}{10} \times \dfrac{4}{49}$

4. $\dfrac{3}{21} \times \dfrac{5}{7}$

5. $\dfrac{3}{5} \times \dfrac{5}{17}$

SECTION 5.2 DIVIDING FRACTIONS

Dividing fractions is just as easy as multiplying them. To divide, multiply the first fraction (dividend) by the reciprocal of the second fraction (divisor) and then simplify.

Recall that you flip the fraction to get its reciprocal. So the steps to dividing fractions are:

- Flip the divisor.
- Multiply fractions.
- Simplify, if possible.

EXAMPLE 1: $\dfrac{2}{3} \div \dfrac{3}{4}$

STEP 1: $\dfrac{2}{3} \times \dfrac{4}{3}$

STEP 1: Flip the divisor (second fraction).

STEP 2: $\dfrac{2 \times 4}{3 \times 3} = \dfrac{8}{9}$

STEP 2: Multiply fractions.

Let's see why we flip the second fraction:

$$\frac{2/3}{3/4} \times \frac{4/3}{4/3} = \frac{8/9}{12/12}$$

Dividing fractions is much easier when we can get rid of the fraction in the denominator. To do this, we multiply both top and bottom by the reciprocal of the denominator. The denominator multiplied by its reciprocal cancels out to make 1. So we are "getting rid" of the fraction in the denominator.

$$\frac{2/3 \times 4/3}{1} = \frac{2}{3} \times \frac{4}{3}$$

So when we say to flip the second fraction (divisor) we are simplifying the process just described.

EXAMPLE 2: $\dfrac{2}{9} \div \dfrac{2}{3}$

STEP 1: $\dfrac{2}{9} \times \dfrac{3}{2}$ **STEP 1:** Flip divisor.

STEP 2: $\dfrac{2 \times 3}{9 \times 2} \div \dfrac{3}{3} = \dfrac{2 \times 1}{3 \times 2} =$ **STEP 2:** Simplify and multiply.

STEP 3: $\dfrac{2 \times 1}{3 \times 2} \div \dfrac{2}{2} = \dfrac{1 \times 1}{3 \times 1}$ **STEP 3:** Simplify and multiply.

EXAMPLE 3:   **Divide** $\dfrac{28}{29}$ **by** $\dfrac{21}{22}$

Remember that the word *by* replaces the division symbol.

$$\frac{28}{29} \times \frac{22}{21}$$

$$\frac{28 \times 22}{29 \times 21} \div \frac{7}{7} = \frac{4 \times 22}{29 \times 3}$$

In this example, the resulting fraction is top heavy, which means that the numerator is larger than the denominator. When this happens, we have to change the answer into a mixed number.

$$\frac{88}{87} = 1\frac{1}{87}$$

 **If you don't remember how to change an improper fraction (sometimes called a top-heavy fraction) to a mixed number, you should review Section 4.**

In the next example, we will be dividing mixed numbers with whole numbers and fractions.

EXAMPLE 4: $\quad \dfrac{4}{7} \div 2\dfrac{2}{3}$

STEP 1: $\dfrac{4}{7} \div \dfrac{8}{3}$

STEP 1: Change the divisor into an improper fraction.

STEP 2: $\dfrac{4}{7} \times \dfrac{3}{8} = \dfrac{4 \times 3}{7 \times 8}$

STEP 2: Flip the divisor and multiply.

STEP 3: $\dfrac{4 \times 3}{7 \times 8} \div \dfrac{4}{4} = \dfrac{1 \times 3}{7 \times 2}$

STEP 3: Simplify.

$\qquad = \dfrac{3}{14}$

PRACTICE:

Show all work. Change improper fractions to mixed numbers.

1. $\dfrac{7}{10} \div \dfrac{1}{2}$

2. $\dfrac{25}{28} \div \dfrac{15}{16}$

3. $6 \div 1\dfrac{1}{8}$

4. $\dfrac{7}{56} \div \dfrac{5}{32}$

5. $5\dfrac{1}{4} \div 3$

EXERCISES

Section 5.1 Multiplying fractions

Simplify and change all improper fractions to mixed numbers.

1. $\dfrac{3}{4} \times \dfrac{2}{15}$

2. $\dfrac{4}{7} \times \dfrac{1}{5}$

3. $\dfrac{5}{6} \times \dfrac{2}{5}$

4. $\dfrac{3}{4} \times \dfrac{1}{21}$

5. $\dfrac{20}{21} \times \dfrac{7}{12}$

6. $\dfrac{5}{6} \times 1\dfrac{1}{3}$

7. $\dfrac{5}{18} \times \dfrac{9}{10}$

8. $2 \times \dfrac{2}{8}$

9. $\dfrac{27}{28} \times \dfrac{2}{9}$

10. $1\dfrac{4}{7} \times 1\dfrac{3}{4}$

11. $\dfrac{7}{8} \times \dfrac{2}{35}$

12. $3\dfrac{1}{3} \times \dfrac{1}{5}$

13. $\dfrac{7}{15} \times \dfrac{1}{7}$

14. $5 \times \dfrac{2}{3}$

Section 5.2 Dividing fractions

Simplify and change all improper fractions to mixed numbers.

15. $\dfrac{2}{12} \div \dfrac{10}{28}$

16. $\dfrac{1}{8} \div \dfrac{3}{4}$

17. $\dfrac{42}{44} \div \dfrac{14}{16}$

18. $\dfrac{6}{10} \div \dfrac{24}{35}$

19. $\dfrac{1}{5} \div \dfrac{2}{5}$

20. $\dfrac{3}{4} \div \dfrac{3}{8}$

21. $\dfrac{28}{29} \div \dfrac{21}{22}$

22. $\dfrac{1}{8} \div 5\dfrac{1}{4}$

23. $\dfrac{35}{36} \div \dfrac{21}{24}$

24. $\dfrac{7}{12} \div \dfrac{5}{9}$

25. $2\dfrac{6}{7} \div 3\dfrac{1}{3}$

26. $\dfrac{1}{8} \div 9$

27. $\dfrac{7}{20} \div \dfrac{7}{10}$

28. $\dfrac{5}{21} \div 1\dfrac{1}{14}$

29. $\dfrac{6}{7} \div \dfrac{12}{42}$

30. $\dfrac{8}{35} \div \dfrac{4}{5}$

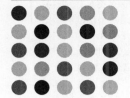

SECTION 6

Fraction and Decimal Conversions and Writing Percents

6.1 CONVERTING FRACTIONS TO DECIMALS

The purpose of this section is to teach you how to convert fractions to decimals. This section will prepare you for question 16 on the Arithmetic Skills Test. When given a fraction it is important to understand that it can also be written as a decimal. Converting between fractions and decimals is essential in some everyday tasks. For instance, when cooking, the recipe you are using may call for $\frac{1}{4}$ pounds of ground beef, and the package of beef that you will use for the recipe is marked as .25 pounds. Therefore, to determine whether the package contains the right amount of meat for the recipe, you can change $\frac{1}{4}$ to .25 pounds.

However, before we start converting fractions to decimals it is important to remember how to divide two numbers, which was covered in Section 3. In addition, it is essential that you remember fraction notation, which was covered in Section 4.

 **If you don't remember, then right now you should go back to Sections 3 and 4 and review division of whole numbers and fraction notation.**

Now that you've reviewed Sections 3 and 4, let's get started! Besides the Arithmetic Skills Test, you will also have to convert fractions to decimals on a daily basis, just like in our recipe example. That is why it is very important to know how to compute fraction conversions. The first step in converting between fractions and decimals is to understand what the fraction is saying. For example, let's convert $\frac{5}{17}$ to a decimal. Our first step in doing so will be to interpret the fraction bar to mean "divided by." Thus $\frac{5}{17}$ is saying in words: 5 divided by 17.

Let's see how $\frac{5}{17}$ will look as a decimal.

STEP 1:

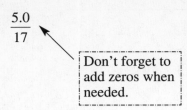

Don't forget to add zeros when needed.

STEP1: Make sure that the numerator has the same amount of digits as the denominator. In this example 17 has two digits. 5 needs one more digit, so we add a zero after its decimal point: 5.0. Now this 17 can be divided into 5.0 without any problems.

STEP 2:

```
     .29
17)5.00
   -3 4
    1 60
   -1 53
       7
```

Dividing the numerator by the denominator

STEP 2: Do what the fraction bar said—divide 5.0 by 17. While dividing you may have to add more zeros after the decimal point. You should also keep dividing until you have a remainder of zero. However, if the directions tell you to give the answer in a specific place value, then stop and round to that place.

STEP 3:

```
     .29 ≈ .3    R 7
17)5.00
  -34
   160
  -153
      7
```

STEP 3: Interpret the information. In this example 5.0 divided by 17 written as a decimal to the hundredths place is 0.29. If we wanted a decimal to the tenths place, then our answer would be 0.3. However, unless advised otherwise, give you answer at least to the hundredths place.

$\frac{5}{17}$ written as a decimal to the hundredths place is 0.29.

Note: It is important to understand that fractions represent the same things as decimals: numbers that are not whole numbers.

Important: Sometimes the division will not end. We call this a **repeating decimal**.

For example $\frac{2}{3} = $

```
   .666...
3)2.000
```

This is a *repeating decimal*, which is the result of the division never ending. Thus, if the division does not end, we must round the quotient. Therefore, $\frac{2}{3}$ is approximately .67.

EXAMPLE 1: Change $\frac{12}{25}$ to a decimal in the hundredths place

STEP 1:

$$\frac{12.00}{25}$$

By doing this our answer will also be a decimal in the hundredths place.

STEP 1: Make sure that the numerator has the same amount of digits as the denominator. In this example, they have the same number of digits. However, the directions tell us to produce a decimal in the hundredths place. So to make sure that will happen, we need to add two zeros to the numerator so that it is a decimal in the hundredths place.

STEP 2:

$$\begin{array}{r} .48 \\ 25\overline{)12.00} \\ -10\,0 \\ \hline 2\,00 \\ -2\,00 \\ \hline 0 \end{array}$$

STEP 2: Do what the fraction bar said: divide 12.00 by 25. Since we have already added the zeros after the decimal point, **NO** additional zeros need to be added!

STEP 3:

$$\begin{array}{r} .48 \\ 25\overline{)12.00} \\ -10\,0 \\ \hline 2\,00 \\ -2\,00 \\ \hline 0 \end{array}$$

STEP 3: Interpret the information. In this example, 12.00 divided by 25 written as a decimal to the hundredths place is 0.48.

Therefore, $\frac{12}{25}$ = .48 .

EXAMPLE 2: Change $\frac{167}{250}$ to a decimal

STEP 1:

$$\frac{167.00}{250}$$

By doing this our answer will also be a decimal in the hundredths place. However, you may have to ADD MORE ZEROS LATER.

STEP 1: Make sure that the numerator has the same amount of digits as the denominator. In this example, they have the same number of digits. However, we want our answer to be in the hundredths place *at least*. So to make sure that will happen, we need to add two zeros to the numerator so that it is a decimal in the hundredths place.

STEP 2:

```
        .668
250) 167.000
    −1500
     1700
    −1500
     2000
    −2000
        0
```

STEP 2: Do what the fraction bar said: divide 167.00 by 250. Now, since the directions didn't specify about place value we can add as many zeros as we want after the decimal point. In this example, if we add one more zero then we will have a remainder of zero.

> Don't stop here! 200 is divisible by 25. So if we add a zero then we can get 2000, which is divisible by 250.

STEP 3:

```
        .668
250) 167.000
    −1500
     1700
    −1500
     2000
    −2000
        0
```

STEP 3: Interpret the information. In this example 167.00 divided by 250 written as a decimal to the thousandths place is 0.668.

Therefore, $\frac{167}{250} = .668$.

PRACTICE:

Convert all fractions to decimals.

1. $\frac{1}{5} =$

2. $\frac{3}{16} =$

3. $\frac{20}{3} =$

4. $\frac{13}{10} =$

6.2 CONVERTING DECIMALS TO FRACTIONS

The purpose of this section is to teach you how to convert decimals to fractions. This section will prepare you for question 15 on the Arithmetic Skills Test. When given a decimal it is important to understand what type of decimal it is. This is because any **terminating decimal** can be written as a fraction. **Terminating** means that the decimal ends. For example, .75 is a decimal that ends; thus it can be written as the fraction $\frac{3}{4}$.

Now, on the other hand, a **nonterminating** and **nonrepeating** decimal **CANNOT** be converted to a fraction. This is because it is an irrational number. *Irrational* means nonfractional. Converting between decimals and fractions is essential in some everyday tasks. For instance, when cooking, the recipe you are using may call for .75 pounds of ground beef and the package of beef that you will use for the recipe is $\frac{1}{4}$ pounds. Therefore, to determine whether the package contains the right amount of meat for the recipe, you can change .25 to $\frac{1}{4}$ pounds.

However, before we start dealing with converting decimals to fractions it is important that you remember the place values of a decimal, which was covered in Section 1. In addition, you must remember how to simplify a fraction into lowest terms and change an improper fraction to a mixed number, which was covered in Section 4.

 **If you don't remember, then right now you should go back to Sections 1 and 4—place values of decimals, simplifying fractions, and changing improper fractions to mixed numbers.**

Now that you've reviewed Sections 1 and 4, let's get started! Besides the Arithmetic Skills Test, you will also have to convert decimals to fractions on a daily basis, just like in our recipe example. That is why it is very important that you know how to compute decimal conversions. The first step in converting between decimals and fractions is to identify what type of decimal it is. You will have four choices:

1. Terminating decimal ◄——— **can change to a fraction**
2. Nonterminating decimal
3. Repeating decimal ◄——— **can change to a fraction**
4. Nonrepeating decimal

STEP 1: .46

Terminating decimal

Let's try to convert .46 to a fraction in lowest terms. **STEP 1:** Identify what type of decimal it is. In this example we have a **terminating decimal**. Therefore, we can convert it into a fraction.

STEP 2: .46

Hundredths place value

STEP 2: Identify what place value the decimal is in.

STEP 3: $\frac{\textcircled{46}}{100}$ ◄— Digits of the decimal

The 100 is equivalent to the hundredths place value

STEP 3: Any terminating decimal can be converted to a fraction by putting the **decimal's digits** over the number that is equivalent to the **place value** of the decimal.

STEP 4: $\dfrac{46}{100} = \dfrac{23}{50}$

STEP 4: Simplify the fraction. Write it in lowest terms.

EXAMPLE 1:   Convert 10.2 to a fraction

Remember to change any improper fractions to mixed numbers.

STEP 1:   10.2

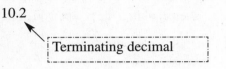

Terminating decimal

STEP 1: Identify what type of decimal it is. In this example we have a **terminating decimal**. Therefore, we can convert it into a fraction.

STEP 2:   10.2

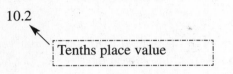

Tenths place value

STEP 2: Identify what place value the decimal is in.

STEP 3: $\dfrac{102}{10}$

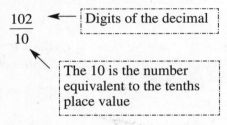

Digits of the decimal

The 10 is the number equivalent to the tenths place value

STEP 3: Any terminating decimal can be converted to a fraction by putting the **decimal's digits** over the number that is equivalent to the **place value** of the decimal.

STEP 4: $\dfrac{102}{10} = \dfrac{51}{5}$

STEP 4: Simplify the fraction. Write it in lowest terms.

STEP 5: $\dfrac{51}{5} = 10\dfrac{1}{5}$

STEP 5: Change the improper fraction to a mixed number.

EXAMPLE 2: Convert 0.577777... to a fraction

Remember to change any improper fractions to mixed numbers.

STEP 1:   0.57777...

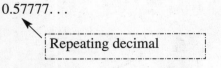

Repeating decimal

STEP 1: Identify what type of decimal it is. In this example we have a **repeating decimal**. Therefore, we can convert it into a fraction.

STEP 2: $x = 0.57777...$

STEP 2: The number 0.57777 is equal to some fraction. Call this fraction "x." **Therefore, set the repeating decimal equal to "x."**

STEP 3:　　　$10x = 5.77777$ p

STEP 3: Since there is only **ONE** repeating decimal, we will multiply both sides by 10. **If there had be three repeating decimals we would have multiplied the sides by 1000.** The number of repeating decimals equals the number of zeros you will have after the 1.

STEP 4:　　　$10x = 5.57777\dots$
　　　　　　　　$\underline{-x = 0.57777\dots}$
　　　　　　　　$9x = 5.20000\dots$

STEP 4: Subtract the original equation from the new equation.

STEP 5:　　　$9x = \dfrac{52}{10} = \dfrac{26}{5}$

　　　　　　　　$x = \dfrac{26}{5} \times \dfrac{1}{9} = \dfrac{26}{45}$

STEP 5: Convert the decimal 5.2 to a fraction and simplify it to lowest terms. Next, solve the equation for "x."

Therefore, $\dfrac{26}{45} = 0.577777$ p

PRACTICE:

Convert all to decimals to fractions in lowest terms.

1. $1.5 =$

2. $0.1666\dots =$

3. $0.078 =$

4. $0.4123123123123\dots =$

6.3 PERCENTS

The purpose of this section is to show you how to find the percent of a number and also identify the relationship between two numbers using percents. This section will prepare you for questions 17 and 18 on the Arithmetic Skills Test. Percents are everywhere in the world today. Whenever you want to buy a piece of clothing on sale you will have to figure out how much of a sale you are getting. For example, say you want to buy a sweater that is marked 30 percnt off of the original price. If the original price is $12.50 how much will you have to pay? The answer is $12.50 − $3.75 = $8.75, or you could have just taken 70 percnt of $12.50. Therefore, it is important to know how to calculate the percent of numbers.

However, before we start dealing with percents it is important that you remember how to multiply fractions, which was covered in Section 5. In addition, you must remember how to simplify a fraction into lowest terms, which was covered in Section 4.

 **If you don't remember, then right now you should go back to Sections 4 and 5 and review simplifying fractions and fraction multiplication.**

Now that you've reviewed Sections 4 and 5, let's get started! The fist step in finding the percent of a number is to change all numbers in percent notation to a fraction.

Important: All numbers in percent notation can be written as a fraction. All you do is take the number that is in percent notation and put it over 100.

For example: $13\% = \dfrac{13}{100}$

Number in percent notation

Let's find 25% of 120.

STEP 1:   25% of 120

> Means multiply

STEP 1: Translate the problem into math language. Remember that "of" in math means "times" or "to multiply." Therefore, the statement reads 25 percent "times" 120.

STEP 2: $\dfrac{25}{100} \times 120$

STEP 2: Write the equation in mathematical notation. Convert the percent to a fraction and substitute a multiplication sign for the word "of."

STEP 3: $\dfrac{25}{\underset{5}{\cancel{100}}} \times \overset{6}{\cancel{120}} = \dfrac{25 \times 6}{5} = \dfrac{150}{5} = 30$

STEP 3: Multiply the fraction times the whole number and simplify whenever possible.

STEP 4:   25% of 120 is 30.

STEP 4: Interpret the information.

Now, what if we want to find out what percent one number is of another number? Well, we perform a similar process. The only difference is that now our answer will actually be a percent.

Percent: 30 is what percent of 120?

STEP 1:   30 is what percent of 120?

STEP 1: Translate the problem into math language. Remember that "of" in math means "times" or "to multiply." Therefore, the statement reads 30 equals *what percent* "times" 120.

STEP 2: $30 = \dfrac{x}{100} \times 120$

STEP 2: Write the equation in mathematical notation. Convert the percent to a fraction and substitute a multiplication sign for the word "of" and an equal sign for the word "is". In addition, for the "what percent" we will use the variable "*x*" to denote our percent.

STEP 3: $\dfrac{30}{120} = \dfrac{x}{100}$

$$\dfrac{1}{4} = \dfrac{x}{100}$$

$$x = 25$$

STEP 3: Solve the equation for x and simplify whenever necessary.

STEP 4:   30 is 25 percent of 120

STEP 4: Interpret the information.

EXAMPLE 1:   Find 16% of 63

STEP 1:   16% of 63

STEP 1: Translate the problem into math language. Remember that "of" in math means "times" or "to multiply." Therefore, the statement reads 16 percent "times" 63.

STEP 2: $\dfrac{16}{100} \times 120$

STEP 2: Write the equation in mathematical notation. Convert the percent to a fraction and substitute a multiplication sign for the word "of."

STEP 3: $\dfrac{\overset{4}{\cancel{16}}}{\underset{25}{\cancel{100}}} \times 63 = \dfrac{4 \times 63}{25} = \dfrac{252}{25}$

$$= 10\dfrac{2}{25}$$

STEP 3: Multiply the fraction times the whole number and simplify whenever possible. Change the improper fraction to a mixed number.

STEP 4:   16% of 63 is $10\dfrac{2}{25}$

STEP 4: Interpret the information.

EXAMPLE 2:   72 is what percent of 96?

STEP 1:   72 is what percent of 96?

STEP 1: Translate the problem into math language. Remember that "of" in math means "times" or "to multiply." Therefore, the statement reads 72 equals *what percent* "times" 96.

STEP 2: $72 = \dfrac{x}{100} \times 96$

STEP 2: Write the equation in mathematical notation. Convert the percent to a fraction and substitute a multiplication sign for the word "of" and an equal sign for the word "is". In addition, for the "what percent" we will use the variable "x" to denote our percent.

STEP 3: $\dfrac{72}{96} = \dfrac{x}{100}$

$$\dfrac{3}{4} = \dfrac{x}{100}$$

$$x = 75$$

STEP 3: Solve the equation for x and simplify whenever necessary.

STEP 4:   72 is 75 percent of 96

STEP 4: Interpret the information.

PRACTICE:

1. Find: 1% of 85 =

2. Find: 25% of 44 =

3. Percent: 1.38 is what percent of 69?

4. Percent: .3 is what percent of 1?

EXERCISES

Section 6.1 Converting fractions to decimals

For exercises 1–12, convert the fraction to a decimal number. Show all work.

Convert.

1. 83/250 to a decimal number =

2. 13/20 to a decimal number =

3. 2/5 to a decimal number =

4. 9/40 to a decimal number =

5. 40/100 to a decimal number =

6. 10/40 to a decimal number =

7. 97/125 to a decimal number =

8. 28/200 to a decimal number =

9. 140/200 to a decimal number =

10. 337/500 to a decimal number =

11. 67/100 to a decimal number =

12. 39/100 to a decimal number =

Section 6.2 Converting decimals to fractions

For exercises 1-12, convert the decimal number to a fraction. Show all work.

Convert.

1. .8 to a fraction in lowest terms =

2. 1.02 to a fraction in lowest terms =

3. .08 to a fraction in lowest terms =

4. 1.2 to a fraction in lowest terms =

5. .2 to a fraction in lowest terms =

6. 1.4 to a fraction in lowest terms =

7. .297 to a fraction in lowest terms =

8. .044 to a fraction in lowest terms =

9. 1.57 to a fraction in lowest terms =

10. .5 to a fraction in lowest terms =

11. .25 to a fraction in lowest terms =

12. .88 to a fraction in lowest terms =

Section 6.3 Percents

For exercises 1–12, find the percents. Show all work.

1. 78% of 81 =

7. 96% of 36 =

2. .005 is what percent of 5?

8. 26% of 93 =

3. 2% of 59 =

9. 10.12 is what percent of 23?

4. 47.5 is what percent of 50?

10. 19% of 49 =

5. 49% of 47 =

11. .546 is what percent of 39?

6. .322 is what percent of 46?

12. 68% of 28 =

13. 475.8 is what percent of 61?

15. 75% of 102 =

14. 8.58 is what percent of 39?

ANSWERS TO PRACTICE PROBLEMS

SECTION 1 – Practice Problems

Section 1.1
1. 95,652

Millions	Hundred thousands	Ten thousands	Thousands	Hundreds	Tens	Ones
0	0	9	5	6	5	2

2. 305,050

Section 1.2
1.

Thousands	Hundreds	Tens	Ones	Tenths	Hundredths	Thousandths	Ten-thousandths	Hundred thousandths
2	5	6	0	2	3	5	6	

5 is in the ten-thousandths place.
2. A. 7 is in the hundredths place.
   B. 2 is in the ten-thousandths place.
3. A. 0.056 < 0.560
   B. 12.010 > 12.0010

Section 1.3
1. Fifty-six hundredths
2. Eight and twenty-five thousandths
3. 10.032
4. 0.0078
5. D; 2.005

Section 1.4
1. 12.8$\underline{7}$5 rounded to the tenths = 12.9
2. 765.22$\underline{5}$ rounded to the hundredths = 765.23
3. 5$\underline{4}$9.006 rounded to the hundreds = 500

SECTION 2 — Practice Problems

Section 2.1

1.
```
     1 1
   8907
   3837
 +   45
  12789
```

2.
```
    1111
  389377
 +  3738
  393115
```

3.
```
   111
  38930
    849
     37
 +   23
  39839
```

4.
```
    1
   937
   390
 +   6
  1323
```

Section 2.2

1.
```
       6
   5 7̶ ¹3 7
 −   6 4 5
   5 0 9 2
```

2.
```
         3
   3 7 4̶ ¹6
 −   1 3 8
   3 6 0 8
```

3.
```
   45578
 −   456
   45122
```

4.
```
   169
 −  48
   121
```

Section 2.3

1.
```
      81.69
      23.24
  +    9.57
     114.50
```

2.
```
      5.99
  +    .86
      6.85
```

3.
```
      52.83
       1.95
  +     .678
      55.458
```

4.
```
      87.2
      15.2
  +    1.3
     103.7
```

Section 2.4

1.
```
      7.29
  −   5.36
      1.93
```

2.
```
      56.95
  −     .869
      56.081
```

3.
```
      32.83
  −   17.95
      14.88
```

4.
```
      87.2
  −   15.2
      72.0
```

SECTION 3 — Practice Problems

Section 3.1

1.
```
        3
       57
  ×    50          57 × 0 = 0
     2850          57 × 5 = 285
```

2.
$$\begin{array}{r} 3 \\ 25 \\ \times\ \underline{17} \\ 175 \\ +\ \underline{250} \\ 425 \end{array}$$
$25 \times 7 = 175$
$25 \times 1 = 25$

3.
$$\begin{array}{r} 2 \\ 47 \\ \times\ \underline{33} \\ 141 \\ +\ \underline{1410} \\ 1551 \end{array}$$
$47 \times 3 = 141$
$47 \times 3 = 141$

4.
$$\begin{array}{r} 65 \\ \times\ \underline{11} \\ 65 \\ +\ \underline{650} \\ 715 \end{array}$$
$65 \times 1 = 65$
$65 \times 1 = 65$

5.
$$\begin{array}{r} 1\ 2 \\ 27 \\ \times\ \underline{23} \\ 81 \\ +\ \underline{540} \\ 621 \end{array}$$
$27 \times 3 = 81$ carry 2
$27 \times 2 = 54$ carry 1

Section 3.2

1.
$$\begin{array}{r} 30.05 \\ \underline{.05} \\ 15025 \\ +\ \underline{00000} \\ 1.5025 \end{array}$$
2 places
2 places

4 places

2.
$$\begin{array}{r} .42 \\ \times\ \underline{.08} \\ 336 \\ +\ \underline{000} \\ .0336 \end{array}$$
2 places
2 places

4 places

3.
$$\begin{array}{r} 12.808 \\ \times\ \underline{7.19} \\ 115272 \\ 128080 \\ +\ \underline{8965600} \\ 92.08952 \end{array}$$
3 places   Line up digits on the right
2 places

5 places

4.
```
        .0075      4 places
  ×     .0510      4 places
        0000
        00750
       037500
 +   0000000
    .00038250      8 places—add extra 0
```

5.
```
        7.82       2 places
  ×     1.03       2 places
        2346
        0000
  +    78200
       8.0546      4 places
```

Section 3.3

1.
```
        22
  66)1452          Check:        66
     −132                     ×   22
      132                        132
     −132                   +   1320
       00                       1452
```

2.
```
        618
  55)34007 r 17    Check:       618
   − 330                     ×   55
     100                       3090
     −55                   +  30900
     457                      33990
    −440                     +   17
      17                     34007
```

3.
```
      10001
  58)580058        Check:     10001
    −58                     ×   58
    000058                   80008
      −58                 + 500050
       00                   580058
```

4.
$$
\begin{array}{r}
2101 \\
32\overline{)67238} \quad r\,6 \\
-64 \\
\hline
32 \\
-32 \\
\hline
0038 \\
-32 \\
\hline
6
\end{array}
$$

Check:
$$
\begin{array}{r}
2101 \\
\times\;32 \\
\hline
4202 \\
+\;63030 \\
\hline
67232 \\
+\;6 \\
\hline
67238
\end{array}
$$

5.
$$
\begin{array}{r}
191 \\
18\overline{)3450} \quad r\,12 \\
-18 \\
\hline
165 \\
-162 \\
\hline
30 \\
-18 \\
\hline
12
\end{array}
$$

Check:
$$
\begin{array}{r}
191 \\
\times\;18 \\
\hline
1528 \\
+\;1910 \\
\hline
3438 \\
+\;12 \\
\hline
3450
\end{array}
$$

Section 3.4

1.
$$
\begin{array}{r}
3 \\
405\overline{)1267} \\
-1215 \\
\hline
52
\end{array}
$$

3.
$$
\begin{array}{r}
.13926 \\
126\overline{)17.54761} \\
-126 \\
\hline
494 \\
-378 \\
\hline
1167 \\
-1134 \\
\hline
336 \\
-252 \\
\hline
841 \\
-746 \\
\hline
85
\end{array}
$$

2.
$$
\begin{array}{r}
.08546 \\
76\overline{)6.49532} \\
-608 \\
\hline
415 \\
-380 \\
\hline
353 \\
-304 \\
\hline
492 \\
-456 \\
\hline
36
\end{array}
$$

4.
$$
\begin{array}{r}
.6470 \\
15\overline{)9.7051} \\
-9\,0 \\
\hline
70 \\
-60 \\
\hline
105 \\
-105 \\
\hline
01
\end{array}
$$

$$
\begin{array}{r}
8.577 \\
630\overline{)5403.889} \\
-5040 \\
\hline
3638 \\
-3150 \\
\hline
4888 \\
-4410 \\
\hline
4789 \\
-4410 \\
\hline
379
\end{array}
$$

5.

SECTION 4 — Practice Problems

Section 4.1

1. $\dfrac{5}{6} + \dfrac{1}{4} = \dfrac{5 \otimes 2}{12} + \dfrac{1 \otimes 3}{12} = \dfrac{10}{12} + \dfrac{3}{12} = \dfrac{13}{12} = 1\dfrac{1}{12}$

2. $\dfrac{1}{5} + \dfrac{3}{10} = \dfrac{1 \otimes 2}{10} + \dfrac{3 \otimes 1}{10} = \dfrac{2}{10} + \dfrac{3}{10} = \dfrac{5}{10} = \dfrac{1}{2}$

3. $\dfrac{31}{50} + \dfrac{9}{50} = \dfrac{31 + 9}{50} = \dfrac{40}{50} = \dfrac{4}{5}$

4. $\dfrac{1}{2} + \dfrac{11}{16} = \dfrac{1 \otimes 8}{16} + \dfrac{11 \otimes 1}{16} = \dfrac{8}{16} + \dfrac{11}{16} = \dfrac{19}{16} = 1\dfrac{3}{16}$

Section 4.2

1. $\dfrac{2}{3} - \dfrac{5}{12} = \dfrac{2 \otimes 4}{12} - \dfrac{5 \otimes 1}{12} = \dfrac{8}{12} - \dfrac{5}{12} = \dfrac{3}{12} = \dfrac{1}{4}$

2. $\dfrac{3}{8} - \dfrac{1}{3} = \dfrac{3 \otimes 3}{24} - \dfrac{1 \otimes 8}{24} = \dfrac{9}{24} - \dfrac{8}{24} = \dfrac{1}{24}$

3. $\dfrac{4}{5} - \dfrac{2}{3} = \dfrac{4 \otimes 3}{15} - \dfrac{2 \otimes 5}{15} = \dfrac{12}{15} - \dfrac{10}{15} = \dfrac{2}{15}$

4. $\dfrac{5}{7} - \dfrac{1}{3} = \dfrac{5 \otimes 3}{21} - \dfrac{1 \otimes 7}{21} = \dfrac{15}{21} - \dfrac{7}{21} = \dfrac{8}{21}$

Section 4.3

1. $4\dfrac{2}{3} + 9\dfrac{5}{6} = 4 + 9 + \dfrac{2}{3} + \dfrac{5}{6} = 13 + \dfrac{4 + 5}{6} = 13 + \dfrac{9}{6} = 13 + 1\dfrac{1}{2} = 14\dfrac{1}{2}$

2. $3\dfrac{1}{2} + 4\dfrac{1}{4} = 3 + 4 + \dfrac{1}{2} + \dfrac{1}{4} = 7 + \dfrac{2 + 1}{4} = 7 + \dfrac{3}{4} = 7\dfrac{3}{4}$

3. $9\frac{1}{10} + 3\frac{1}{2} = 9 + 3 + \frac{1}{10} + \frac{1}{2} = 12 + \frac{1+5}{10} = 12 + \frac{6}{10} = 12\frac{3}{5}$

4. $2\frac{7}{8} + 1\frac{1}{2} = 2 + 1 + \frac{7}{8} + \frac{1}{2} = 3 + \frac{7+4}{8} = 3 + \frac{11}{8} = 3 + 1\frac{3}{8} = 4\frac{3}{8}$

SECTION 5 — Practice Problems

Section 5.1

1. $\frac{4 \times 1}{7 \times 3} = \frac{4}{21}$

2. $\frac{13 \times 8}{16 \times 52} \div \frac{8}{8} = \frac{13 \times 1}{2 \times 52}$

   $\frac{13 \times 1}{2 \times 52} \div \frac{13}{13} = \frac{1 \times 1}{2 \times 4} = \frac{1}{8}$

3. $\frac{7 \times 4}{10 \times 49} \div \frac{7}{7} = \frac{1 \times 4}{10 \times 7}$

   $\frac{1 \times 4}{10 \times 7} \div \frac{2}{2} = \frac{1 \times 2}{5 \times 7} = \frac{2}{35}$

4. $\frac{3 \times 5}{21 \times 7} \div \frac{3}{3} = \frac{1 \times 5}{7 \times 7} = \frac{5}{49}$

5. $\frac{3 \times 5}{5 \times 17} \div \frac{5}{5} = \frac{3 \times 1}{1 \times 17} = \frac{3}{17}$

Section 5.2

1. $\frac{7}{10} \times \frac{2}{1} = \frac{7 \times 2}{10}$

   $\frac{7 \times 2}{10} \div \frac{2}{2} = \frac{7}{5} = 1\frac{2}{5}$

2. $\frac{25}{28} \times \frac{16}{15} = \frac{25 \times 16}{28 \times 15} \div \frac{5}{5} = \frac{5 \times 16}{28 \times 3}$

   $= \frac{5 \times 16}{28 \times 3} \div \frac{4}{4} = \frac{5 \times 4}{7 \times 3} = \frac{20}{21}$

3. $\frac{6}{1} \div \frac{9}{8} = \frac{6}{1} \times \frac{8}{9}$

   $= \frac{6 \times 8}{9} \div \frac{3}{3} = \frac{2 \times 8}{3} = 5\frac{1}{3}$

4. $\dfrac{7}{56} \times \dfrac{32}{5} = \dfrac{7 \times 32}{56 \times 5}$

$\qquad = \dfrac{7 \times 32}{56 \times 5} \div \dfrac{8}{8} = \dfrac{7 \times 4}{7 \times 5} = \dfrac{4}{5}$

5. $\dfrac{21}{4} \times \dfrac{1}{3} = \dfrac{21}{4 \times 3}$

$\qquad = \dfrac{21}{4 \times 3} \div \dfrac{3}{3} = \dfrac{7}{4} = 1\dfrac{3}{4}$

SECTION 6 — Practice Problems

Section 6.1

1. $\dfrac{1}{5} = \dfrac{1.00}{5} = 5\overline{)1.00} = .20$

$$
\begin{array}{r}
.20 \\
5\overline{)1.00} \\
-\underline{1.0} \\
00 \\
-\underline{0} \\
0
\end{array}
$$

2. $\dfrac{3}{16} = \dfrac{3.00}{16} = 16\overline{)3.0000} = .1875$

$$
\begin{array}{r}
.1875 \\
16\overline{)3.0000} \\
-\underline{16} \\
140 \\
-\underline{128} \\
120 \\
-\underline{112} \\
80 \\
-\underline{80} \\
0
\end{array}
$$

3. $\dfrac{20}{3} = \dfrac{20.00}{3} = 3\overline{)20.00} = 6.67$

$$
\begin{array}{r}
6.66 \\
3\overline{)20.00} \\
-\underline{18} \\
20 \\
-\underline{18} \\
20 \\
-\underline{18}
\end{array}
$$

4. $\dfrac{13}{10} = \dfrac{13.00}{10} = 10\overline{)13.00} = 1.30$

$$
\begin{array}{r}
1.30 \\
10\overline{)13.00} \\
-\underline{10} \\
30 \\
-\underline{30} \\
0
\end{array}
$$

Section 6.2

1. $1.5 = \dfrac{15}{10} = \dfrac{3}{2}$

2. $0.1666\ldots =$

$$x = 0.1666 \text{ p}$$

$$10x = 1.6666\ldots$$
$$-x = 0.1666\ldots$$
$$\overline{9x = 1.5}$$

$$9x = \dfrac{15}{10} = \dfrac{3}{2}$$

$$x = \dfrac{3}{2} \cdot \dfrac{1}{9} = \dfrac{3}{18}$$

$$x = \dfrac{1}{6}$$

3. $0.078 = \dfrac{78}{1000} = \dfrac{39}{500}$

4. $0.4123123123123 =$
$$x = 0.4123123123\ldots$$

$$1000x = 412.3123123123\ldots$$
$$-x = 0.4123123123\ldots$$
$$\overline{999x = 411.9}$$

$$999x = \dfrac{4119}{10} =$$

$$x = \dfrac{4119}{10} \cdot \dfrac{1}{999} = \dfrac{4119}{9990}$$

$$x = \dfrac{1373}{3330}$$

Section 6.3

1. Find: 1% of 85 =

$$\dfrac{1}{100} \otimes 85 = \dfrac{85}{100} = .85 \quad 1\% \text{ of } 85 \text{ is } .85$$

2. Find: 25% of 44 =

$$\dfrac{25}{100} \otimes 44 = \dfrac{1}{4} \otimes 44 = 11 \quad 25\% \text{ of } 44 \text{ is } 11$$

3. Percent: 1.38 is what percent of 69?

$$\dfrac{138}{100} = \dfrac{x}{100} \otimes 69$$

$$\dfrac{1}{69} \otimes \dfrac{138}{100} = \dfrac{x}{100}$$

$$\frac{2}{100} = \frac{x}{100}$$

$x = 2$ 1.38 is 2 percent of 69

4. Percent: .3 is what percent of 1?

$$.3 = \frac{3}{10} \text{ multiply top and bottom by 10}$$

$$\frac{30}{100} = \frac{x}{100} \otimes 1$$

$x = 30$ .3 is 30 percent of 69

ANSWERS TO SELECTED EXERCISES

Section 1

1.1   Whole Number Place Value
1.   17,430
3.   400,041,000
5.   6,000,052
7.   82,074
9.   400,078

1.2   Decimal Place Value
11.   Hundredths
13.   Ten-thousandths
15.   Tenths
17.   3.0682 > 3.0628

1.3   Decimal Forms
19.   Thirty-eight thousandths
21.   Two and fourteen thousandths
23.   0.02
25.   0.416
27.   0.0090

1.4   Rounding
29.   0.27
31.   1.6
33.   1.601
35.   1.4

Section 3

3.1   Whole Number Multiplication
1.   1,682
3.   682
5.   6,364
7.   2,146
9.   4,355
11.   1,610
13.   4,270

3.2   Decimal Multiplication
15.   0.00009373
17.   0.011336

19. 0.03108
21. 0.002146
23. 0.0005035
25. 0.00495
27. 0.067

3.3   Whole Number Division
29. 10,001
31. 22
33. 60,002
35. 33
37. 40,009
39. 307
41. 73

3.4   Decimal Division
43. 8.8
45. 0.81
47. 0.16
49. 0.65
51. 0.0081
53. 0.0046
55. 0.0049

Section 5

5.1   Multiplying fractions

1. $\dfrac{1}{10}$ 3. $\dfrac{1}{3}$ 5. $\dfrac{5}{9}$

7. $\dfrac{1}{4}$ 9. $\dfrac{3}{14}$ 11. $\dfrac{1}{20}$

13. $\dfrac{1}{15}$

5.2   Dividing Fractions

15. $\dfrac{7}{15}$ 17. $1\dfrac{1}{11}$ 19. $\dfrac{1}{2}$

21. $1\dfrac{1}{87}$ 23. $1\dfrac{1}{9}$ 25. $\dfrac{6}{7}$

27. $\dfrac{1}{2}$ 29. 3